Bayesian Economics
Through
Numerical Methods

Springer
New York
Berlin
Heidelberg
Barcelona
Budapest
Hong Kong
London
Milan
Paris
Santa Clara
Singapore
Tokyo

Jeffrey H. Dorfman

Bayesian Economics Through Numerical Methods

A Guide to Econometrics and Decision-Making with Prior Information

Springer

Jeffrey H. Dorfman
Department of Agricultural
  and Applied Economics
University of Georgia
Athens, GA 30602-7509
USA

Library of Congress Cataloging-in-Publication Data
Dorfman, Jeffrey H.
    Bayesian economics through numerical methods : a guide to
  econometrics and decision-making with prior information / Jeffrey H.
  Dorfman.
        p.   cm.
    Includes bibliographical references and index.
    ISBN 0-387-98233-7 (hc : alk. paper)
    1. Econometrics.   2. Bayesian statistical decision theory.
  I. Title.
  HB139.D674   1997
  330′.01′5195—dc21   97-12147

Printed on acid-free paper.

Production managed by Lesley Poliner; manufacturing supervised by Jacqui Ashri.
Typeset by Bartlett Press from WordPerfect files supplied by the author.
Printed and bound by R.R. Donnelley and Sons, Harrisonburg, VA.
Printed in the United States of America.

9 8 7 6 5 4 3 2 1

ISBN 0-387-98233-7 Springer-Verlag New York Berlin Heidelberg   SPIN 10574679

To Melody,
for too many reasons to list.

Contents

1

Introduction

Bayesian statistics at its most basic level is an approach to statistical problems that seeks to optimally combine information from two sources: the information the researcher believes at the start of the research process and the information contained in the data. Bayes' theorem is essentially a rule for how to combine these two sources of information into a single set of (updated) information concerning the parameters or hypotheses of interest. There are a number of advantages to this approach as opposed to the alternative, sampling theory, approach to statistics.

First, by formalizing the researcher's advance beliefs, through a mathematical construct called the *prior distribution*, underlying assumptions are exposed that can often remain hidden in the sampling theory approach. Second, the use of loss functions to estimate parameters and make optimal decisions concerning hypotheses or settings for control variables allows the statistical process to be customized to fit the particular application. Many other advantages will be discussed in Chapter 2.

The goal of this book is twofold. First, I hope to convince you of the usefulness and inherent advantages of Bayesian statistics relative to the sampling theory, or classical, approach. To do this, basic concepts of Bayesian statistics and decision science will be presented with references given to more complete treatments on these topics. Then, new numerical methods that have been developed in the last 30 years will be covered. These methods have greatly expanded the range of problems to which Bayesian methods can be applied without having to make unrealistic simplifying assumptions. After mastering these numerical methods, researchers can tackle virtually any problem in the area of statistics or decision science within the Bayesian paradigm.

The second goal of this book is to provide a roadmap of applied economic questions that can now be answered empirically with Bayesian methods, emphasizing problems that are best solved with numerical Bayesian methods. Many researchers, unable to keep abreast of all developments in numerical methods, may still think that Bayesian statistics is confined to the arena of problems that can be attacked analytically, forcing simplifications they hope to avoid in applied work. This book hopes to demonstrate the breadth of economic topics that can now be investigated by taking a Bayesian approach, thanks to the advances in numerical techniques. In Chapters 4 through 10, a wide variety of empirical applications will be presented in considerable detail, describing how Bayesian methods have been used to study these problems in the past and how they can be investigated in the future. These chapters will try to provide a guide you can follow in your own research, showing step by step how to apply numerical Bayesian methods to achieve empirical solutions to important economic questions.

Third, because this book hopes to encourage more researchers to "convert" to the Bayesian approach, an extensive bibliography, grouped by category, is included. This list includes references to more complete treatments of the theoretical issues involved as well as a multitude of published empirical studies proving that Bayesian methodology can successfully solve real-world empirical problems in economics. These studies should provide further guidance on the application of these tools to the problems you seek to solve.

The level of knowledge assumed by this book is fairly minimal. Readers are assumed to have an advanced-undergraduate-level grasp of statistics covering such concepts as probability density functions, basic statistical distributions, the likelihood function for a set of observations, hypothesis testing, the central limit theorem, the expectations operator, and a basic familiarity with integral and differential calculus. No prior knowledge or familiarity with Bayesian statistics or computer-based random number generation and simulation techniques is assumed. However, it is assumed that the reader is generally competent at using standard econometrics computer software and can translate the steps laid out in the applications chapters into commands for the econometrics package, statistics software program, or programming language of their choice.

Proofs of the results and extensive mathematical details are not the strength of this book. The references listed in the book contain many proofs and high-level mathematical expositions on the theoretical points involved in these topics, and readers will eventually need to read some of these other sources to gain a complete, expert knowledge of these topics. This book aims to concentrate more on the intuition and implementation of these techniques. It is a user's guide to Bayesian econometrics, not a treatise on the mathematical statistics of the topic. Readers will hopefully finish the book with a greater appreciation for Bayesian methodology, with a basic and intuitive understanding of the concepts involved, and with the ability to apply the techniques described in this book to real-world empirical problems.

Finally, let me thank many of the people who have helped me during the process of becoming a Bayesian and in the writing of this book. The two people who first introduced me to Bayesian statistics are Art Havenner and Arnold Zellner, and they deserve much credit for helping start me on the road to understanding and applying the techniques found in this book. John Geweke was also especially helpful during the process, discussing technical issues with me and providing insight. I have also benefited greatly from conversations about econometrics and decision theory, both Bayesian and classical, with John Antle, Masanao Aoki, David Bessler, Wade Brorsen, Oscar Burt, Jim Chalfant, David Dickey, Ken Foster, Richard Green, Bill Griffiths, Dale Heien, Garth Holloway, Scott Irwin, George Judge, Cathy Kling, Dave Kraybill, Bill Lastrapes, Sergio Lence, Jim LeSage,

David Li, Rob McCulloch, Anya McGuirk, Chris McIntosh, Ron Mittelhammer, John Monahan, Quirino Paris, Peter Phillips, Dale Poirier, Peter Rossi, P. A. V. B. Swamy, Leigh Tesfatsion, Wally Thurman, George Tiao, Herman van Dijk, Mike West, Charles Whiteman, Carl Zulauf, and many other colleagues, some of whom surely belong in this list. None of these colleagues, of course, deserves any blame for what actually appears in the book, although credit for the positive parts should be shared. I owe thanks to Teresa Byrd for help with the figures, bibliography, and so many tasks over the past eight years. Lastly, Martin Gilchrist of Springer-Verlag was an encouraging, supportive editor who decided that I could write a book and allowed me to do so, while Lesley Poliner ably supervised the production process and turned a manuscript into a book.

Part I

Theory and Basics

2

A Quick Course in Bayesian Statistics and Decision Theory

This chapter is not meant to be a complete course in Bayesian statistics and decision theory, but rather an introduction to these topics. For a fuller treatment of these concepts, consult the references listed at the end of this chapter or under "Key Theory Details" in the Bibliography. This chapter hopes to acquaint the reader with the basic vocabulary and ideas that are central to the Bayesian methodology and to teach the standard tools that are applied in most situations.

Bayesian statistics always uses all available information from two sources: the researcher's prior beliefs about the problem being studied and the data collected for the current study. The researcher's prior beliefs can be based on prior studies, theories concerning the subject area, or nonformal observations, but generally should not be based on the data set to be used in the current application.

Bayes' Theorem

At the heart of the Bayesian paradigm lies Bayes' theorem. Take common notation from econometrics and denote data by (y, X), where y is a vector of random observations and X is a matrix of predetermined variables that are used to help explain the variations within y. Further, let θ denote a vector of random parameters that (along with X) control the probability density function of y. In the classical linear regression framework, y is the dependent variable, X is the matrix of independent variables or regressors, and θ would be a vector that includes the regression coefficients and error variance. Bayes' theorem is written as

$$p(\theta|y, X) = \frac{p(\theta)p(y|\theta, X)}{p(y)} \propto p(\theta)p(y|\theta, X) \qquad (2.1)$$

where $p()$ represents a probability density function. The function $p(\theta)$ is the prior distribution on the random parameters θ representing the researcher's *subjective* beliefs about these parameters before analyzing the data contained in (y, X); $p(y|\theta, X)$ is the conditional density of y given a particular set of values for θ and X, commonly referred to as the *likelihood function* of y, and $p(y)$ is often

The information flow of Bayes' theorem

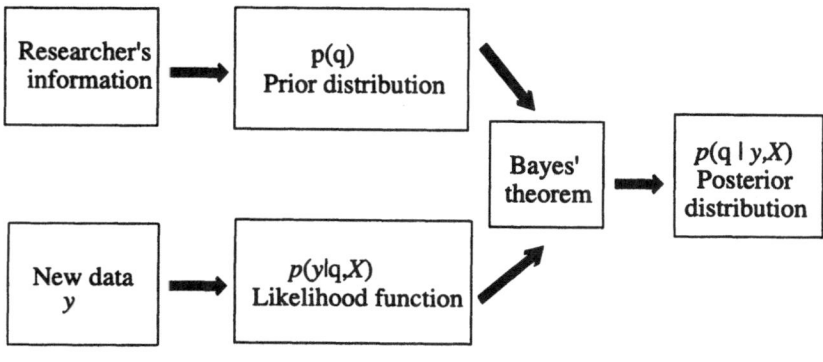

FIGURE 1.

referred to as the *marginal likelihood* of y (because the effect of the independent variables and random parameters has been integrated out). The density $p(\theta|y, X)$ is the posterior distribution of θ representing an updating of the researcher's prior beliefs to account for the new information contained in the data (y, X). At its heart, Bayes' theorem is simply a statistical rule based on the standard theory of probability density functions that shows how to optimally combine information from two sources (the prior and the data). In fact, not only does Bayes' theorem combine these two pieces of information optimally, but any other combination would violate the logical (and mathematical) coherence of the rules concerning the operation of probability distributions.

The first version of Bayes' theorem can be written as an equality due to the denominator of the unconditional density of y, which serves as the normalizing constant to ensure that the resulting posterior conditional density of θ is a proper one and integrates to 1 over the allowable range for the parameters. The second version, holding only to within a constant of proportionality, is missing this normalization, as is commonly seen in the literature. In fact, in empirical work, this second form is the most common. Researchers compute this proportional posterior of θ initially and the normalizing constant later if necessary (in many examples, ratios are taken and the normalizing constant cancels out).

Figure 1 shows schematically how two parallel sources of information are brought together through Bayes' theorem to reach the posterior distribution of the parameters θ. The data and the prior information are brought together through Bayes' theorem to produce the posterior distribution. The relative influence of the two information sources will depend on their relative precision; that is, the smaller the variance of the distribution, the larger its role in the formation of the posterior distribution.

Prior Distributions

The prior distribution describes the researcher's beliefs about the parameters of interest, represented earlier by θ, before gaining any information from the current data set (y, X). Prior distributions are subjective in the sense that because they represent the researcher's beliefs, they cannot be incorrect in a strict sense, but they should still meet basic tests of reasonableness. For example, they should not be so concentrated as to disallow any contribution of information from the data set nor should they be at odds with conventional wisdom without a strong basis for inviting controversy (Box and Tiao, 1973).

Prior distributions are often classified according to some of the following conventions. A prior distribution can be *informative*, conveying prior knowledge concerning at least some elements of θ, or *noninformative*, conveying ignorance. The most famous noninformative, or ignorance, prior is the Jeffreys' prior, which is taken to be proportional to the square root of the determinant of the information matrix (i.e., the negative expected value of the second derivative of the log-likelihood function with respect to the parameters θ). In the classical linear regression model with normally distributed errors, where $\theta = (\beta', \sigma)'$, the Jeffreys' prior is given by $p(\theta) \propto 1/\sigma$. In simple terms, Jeffreys suggested that if the researcher wishes to express prior ignorance, a prior distribution proportional to a constant for parameters with infinite admissable range (i.e., a uniform distribution) should be used and for parameters (such as the standard deviation) that have admissable range of $[0, \infty)$ a uniform distribution prior on the natural log of the parameter should be used (Zellner, 1971, p. 41–53). It is these rules that lead to the prior for the classical regression model given earlier.

The concept of an informative prior can be made concrete using the construct of a simple linear regression model where the dependent variables y are observations on the quantity sold of a product, expressed in natural log format, and the X matrix includes only a constant and the natural log of the price of the product. Further, assume that some miracle allows the error variance to be known. Then θ consists of two regression parameters, denoted here by α and β, respectively, for the intercept and slope coefficients. Due to the logarithmic form of the data, β is the price elasticity of demand and many researchers will have reasonably well-formed prior beliefs of its value, depending on the specific product being modeled. However, less will be known about the intercept, so ignorance may be a reasonable prior belief with respect to α. Thus, for many food products that generally have inelastic demands, one might construct a prior distribution, $p(\alpha, \beta) \propto N[(\beta + 0.5)/0.25]$ where $N[\ ]$ is a standard normal distribution. This prior would place almost all the prior support (± 2 standard deviations) in the inelastic range of $(-1, 0)$, allowing for some slight possibility of an alternative outcome, and centers the prior belief around -0.5. If such a prior distribution is felt by a researcher to be too informative, the variance can be increased; however, caution should be exercised to protect against the prior distribution becoming so diffuse as to place considerable prior support on values that are not truly credible a priori.

Prior distributions can be proper or improper. A *proper* prior distribution is one that integrates to 1 over the admissable range of θ. Distributions that are informative are almost always proper subject to the proper scaling; for example, if a truncated normal distribution is used for a prior distribution on a parameter that has a limited range, the distribution must be scaled to account for the truncation and renormalize the distribution to have total prior support equal to 1. Noninformative priors are often improper. The uniform prior over infinite range is *improper*, as it integrates to infinity, not unity. Improper priors cause problems in the arena of hypothesis testing, where they can cause arbitrary scaling of posterior odds ratio (discussed in depth soon) and make it impossible to assess the relative support for competing models. For this reason, they should be avoided when possible (and suitable). In most applications, a weakly informative prior can be constructed that allows the likelihood function of the data to determine the posterior distribution almost completely while still preventing the problem of arbitrary scaling in odds ratios for hypothesis tests.

Prior distributions should be specific to the situation. In economics, we often have prior information on the sign of parameters, on the relative or approximate magnitude of parameters, and even on (sometimes complex) functions of parameters. For example, when estimating a demand curve in quantity-dependent form, most researcher's would feel they have good prior information that the parameter associated with the good's price should be negative. The prior distribution the researcher constructs should, therefore, incorporate this information; failure to do so would result in a prior distribution that does not fully incorporate all the information actually in the researcher's possession. This would be analogous to inefficient estimation in the classical sampling theory context.

A *conjugate* prior is one that when combined with the likelihood function (and its parametric form) leads to a posterior distribution that has a form that allows analytical inspection (i.e., it has a recognizable form of a known distribution such as the multivariate Student-t distribution). Before recent advances in the theory of numerical methods for analyzing distributions and evaluating integrals, conjugate priors were very important in Bayesian analysis. Researchers had to work hard to conform their prior beliefs to fit within the constraint of the particular distributional class that was conjugate for the assumed likelihood. Given the new advances in numerical techniques for constructing posterior distributions (discussed in depth in the next chapter), conjugate priors have become unnecessary, and researchers are free to choose any prior that truly represents their prior beliefs. In fact, when working with colleagues on Bayesian research I often try to get them to draw a picture of what they think a reasonable prior distribution would look like. Then I use my computer to draw pictures of various distributions (normal, t, beta, etc.), changing the parameters that control the distribution's shape until the colleagues agree that it matches their beliefs. This gives us a mathematical representation of our prior information and is fairly easy to accomplish if at least one collaborator can picture (or look up pictures of) a variety of statistical distributions.

Estimation and Loss Functions

After specifying the prior distribution for the unknown random parameter vector θ, the researcher must specify the form of the likelihood function for the observations y. Generally, the likelihood takes a specific parametric form (often assuming the observations are normally distributed), but the form can be customized to the particular features of the data (limited, qualitative, skewed, etc.) or it even can be nonparametric. The minimum requirement is that the likelihood specified can be evaluated for a given set of data (y, X) and parameter value θ.

Now that a prior distribution and likelihood function are in place, the posterior distribution, $p(\theta|y, X)$, can be constructed, either analytically or numerically. A variety of methods to numerically derive the posterior distribution will be discussed in detail in the next chapter; for now, just assume that the posterior distribution is available in at least empirical form (i.e., we can numerically compute any desired probability statements about θ). For some applications, the desired result is the complete posterior distribution or some probability statement concerning a subset of parameters or function of parameters from θ. In these cases, a point estimate is not required; however, researchers often want a point estimate of θ to present or use for a subsequent purpose.

To derive Bayesian posterior point estimates of unknown random parameters, a researcher must first specify the loss function. A *loss function* measures the loss caused by an estimation error as a function of the parameter estimate and the true (unknown) value. Thus, a Bayesian posterior point estimate for θ is found by solving the optimization problem

$$\text{choose } \hat{\theta} : \quad \min EL(\hat{\theta}, \theta) = \int L(\hat{\theta}, \theta) p(\theta|y, X)\, d\theta. \quad (2.2)$$

The loss function L can take many forms and should be chosen carefully to match the characteristics of the problem being studied. The loss function can be linear or nonlinear in θ and $\hat{\theta}$. Common choices of loss functions are:

a. $L(\hat{\theta}, \theta) = (\hat{\theta} - \theta)^2$ quadratic loss,   (2.3)

b. $L(\hat{\theta}, \theta) = |\hat{\theta} - \theta|$ absolute loss,   (2.4)

c. $L(\hat{\theta}, \theta) = 0$ if $\hat{\theta} = \theta$
            $= c$ if $\hat{\theta} \neq \theta$ zero-one loss.   (2.5)

Choosing a quadratic loss function results in the optimal Bayesian point estimate being the mean of the posterior distribution. The absolute loss function produces a point estimate that equals the median of the posterior distribution. The zero-one loss function yields a point estimate equal to the mode of the posterior distribution. If the posterior distribution is symmetrical, all three of these loss functions will produce the same point estimate of θ. However, for many cases with interesting prior distributions or nonsymmetrical distributions for the observations y, these loss functions will yield different point estimates for θ. Choosing the posterior mode (the most "likely" point in the posterior distribution) implies that even a tiny estimation error is as bad as a huge one. Choosing the posterior median protects against outliers and skewed tails of the distribution and may produce a more robust estimator than the posterior mean. The posterior mean is the only one of these three estimators that uses all the information in the posterior distribution to derive the point estimate.

Specialized loss functions should not be overlooked. For example, when choosing a dimension for the model order of a state-space time series model, Dorfman and Havenner (1992) used an asymmetric quadratic loss function for the model order in the form

$$L(n, \hat{n}) = c(n - \hat{n})^2 I(n - \hat{n}) + (n - \hat{n})^2[1 - I(n - \hat{n})], \qquad (2.6)$$

where I is an indicator function that equals 1 when the argument is positive and 0 when the argument is negative, and c is a scalar greater than 1. Thus, this loss function penalizes estimation errors that result in the model order being too small more severely than those that overparameterize the model. The motivation for such a loss function is that overparameterization leads to some additional (finite sample) samping error in estimation, but underparameterization leads to biased and inconsistent coefficients. This is just one example of a specialized loss function; many such functions that are well-suited to particular applications can, and should, be constructed when warranted.

Hypothesis Testing

Bayesian hypothesis testing is conducted through the mechanism of the *posterior odds ratio*, the ratio of the posterior probability of one hypothesis to the other. For a simple hypothesis such as

$$H_1 : \theta \in \Theta_1 \quad \text{vs.} \quad H_2 : \theta \in \Theta_2, \qquad (2.7)$$

where the two sets Θ_1 and Θ_2 are mutually exclusive and exhaustive, the posterior odds ratio is given by

$$K_{12} = \int_{\Theta_1} p(\theta|y, X) \, d\theta \Big/ \int_{\Theta_2} p(\theta|y, X) \, d\theta = p_1/(1 - p_1) \qquad (2.8)$$

where p_1 is the posterior support for the first hypothesis. As can be seen from equation (2.8), the posterior odds ratio compares the posterior support for one hypothesis to the other. Posterior odds ratios can be constructed for hypotheses that are not mutually exhaustive, in which case the second equality of equation (2.8) obviously would not hold.

In applied research it is often common to also report the Bayes factor, which is the ratio of likelihood support for the two hypotheses. In other words, the Bayes factor is a posterior odds ratio computed with equal prior probabilities for the two competing hypotheses. For the hypotheses described in equation (2.7), the Bayes factor would be given by

$$B_{12} = \int_{\Theta_1} p(y|\theta, X) \, d\theta \Big/ \int_{\Theta_2} p(y|\theta, X) \, d\theta. \qquad (2.9)$$

The benefit of reporting Bayes factors as part of the empirical results of a hypothesis testing exercise is that the effect of the researcher's prior distribution on the test results can then be clearly discerned by comparison of the Bayes factor to the posterior odds ratio. Note that for the posterior odds ratio in (2.8) and the Bayes factor in (2.9) to be well-defined, the prior distributions over θ should be proper, and for the definition of the Bayes factor the prior distributions over the two subsets Θ_1 and Θ_2 should integrate to the same value (e.g., 1).

Conditional, Marginal, and Joint Distributions

This book will discuss probability distributions for vector-valued random variables and will sometimes identify these distributions by one of the types in the section heading. To review very quickly, a joint distribution, $p(\beta, \gamma)$, is for more than one random variable and describes the probabilities of a particular combination of values for the random variables occurring. A conditional distribution describes the probabilities of various values for a (vector-valued) random variable occurring conditional on a specific value of some other random variable whose value is being held constant. Conditional distributions will be written as $p(\beta|\gamma)$. Finally, the marginal distribution for a random variable is its probability distribution without regard for the possible random values of other (potentially correlated) variables. Marginal distributions will be written as $p(\beta)$ and are mathematically related to joint distributions by the formula $p(\beta) = \int p(\beta, \gamma)\, d\gamma$, which shows that the effect of values of the random variable γ have been integrated out to leave only the probabilities of various values of β "averaged" across all possible values for γ.

Some Standard Distributions from Regression Models

Given the prevalence of linear regression models, it is probably worthwhile to present some of the standard distributions that appear in Bayesian analysis of such models under commonly used diffuse prior distributions. While I do not necessarily recommend using such noninformative priors often, they serve as a benchmark and allow for analytical inspection of the posterior distributons of the regression parameters. Further, the distributions presented in this section are frequently useful in the generation of random draws for some of the numerical integration techniques described in the next chapter.

We begin with a standard linear regression model with all the classical assumptions (full rank, nonstochastic regressor matrix, iid zero-meaned, normally distributed error terms), using the notation

$$y = X\beta + \varepsilon, \tag{2.10}$$

where y is an $(n \times 1)$ vector of observations on the dependent variable whose variation the model seeks to explain, X is the regressor matrix, β is a $(k \times 1)$ vector or the regression parameters to be estimated, and ε is the conformable vector of white noise error terms with variance σ^2. If we selected an improper uniform prior for the regression parameters (which is also a Jeffreys' prior) and the standard Jeffreys' prior on σ, the joint prior distribution on (β, σ) can be written as

$$p(\beta, \sigma) \propto 1/\sigma. \tag{2.11}$$

The likelihood function for the observations on the dependent variable given the assumption of normality is

$$p(y|\beta, \sigma, X) = (2\pi\sigma^2)^{-n/2} \exp[-0.5(y - X\beta)'(y - X\beta)/\sigma^2]. \tag{2.12}$$

The joint posterior distribution can then be expressed as

$$p(\beta, \sigma|y, X) \propto (2\pi)^{-n/2}\sigma^{-(n+1)} \exp[-0.5(y - X\beta)'(y - X\beta)/\sigma^2], \tag{2.13}$$

$$\propto (2\pi)^{-n/2}\sigma^{-(n+1)}$$

$$\times \exp[-0.5(n - k)s^2/\sigma^2 + (\beta - \hat{\beta})'\sigma^{-2}X'X(\beta - \hat{\beta})], \tag{2.14}$$

where the expression in (2.14) breaks the quadratic terms apart and uses the least squares estimators of β and σ^2, $\hat{\beta}$ and s^2. The marginal posterior distributions of the parameters are

$$p(\beta|y, X) \propto [(n - k)s^2 + (\beta - \hat{\beta})'X'X(\beta - \hat{\beta})]^{-n/2}, \tag{2.15}$$

which is a multivariate Student-t distribution, and

$$p(\sigma|y, X) \propto \sigma^{-(n-k+1)} \exp[-0.5(n - k)s^2/\sigma^2], \tag{2.16}$$

which is an inverted gamma distribution and implies that the marginal posterior distribution of σ^2 is a scaled version of the χ^2 distribution. When dealing with generalized error distributions with a variance-covariance matrix denoted by Σ, the preceding distributions generalize in obvious ways and the marginal posterior distribution of Σ is an inverted-Wishart distribution, which is a generalized form of the inverse gamma. These distributions will serve as useful references for the remainder of this book.

Comparison to the Sampling Theory Approach

It may be useful for readers who are not particularly familiar with the Bayesian approach to statistics to consider the following four comparisons between Bayesian statistics and sampling theory statistics (the "classical" approach). These comparisons are meant to highlight the differences between the two approaches; they focus mainly on distinctions between numerical Bayesian techniques and sampling theory approaches such as maximum likelihood and least squares estimation.

1 Finite Sample Properties

All Bayesian results are exact in finite samples because the distributions are derived conditional on the observed sample of data. Many sampling theory results depend on asymptotics and are only approximations for the observed sample of data (often poor ones for small to medium samples). This distinction becomes particulary acute when the focus of attention falls on a nonlinear function of structural parameters. In such cases it can be difficult to conduct exact statistical inference in the sampling theoretic framework, and researchers are forced to rely on asymptotic testing procedures, which may tend to overreject the null hypothesis. When taking a numerical Bayesian approach, nonlinear functions of structural parameters pose no problems. Exact distributions of any function of random variables can be constructed easily using the techniques described in the next chapter, allowing inferences concerning any hypotheses of interest.

2 Ease of Computing Precision

In some econometric applications the random variable of key interest to the researcher (or a policymaker who will be using the results as a decision aid) is a nonlinear function of the structural parameters of the regression model. Examples of such cases are the welfare impacts of trade legislation, elasticities of demand and supply, and impulse response functions. Using numerical Bayesian techniques, measures of precision (such as standard errors) for such variables can be computed in as straightforward a manner as for the structural parameters. No additional steps are necessary, and all results are exact in finite samples. Sampling theory approaches generally must rely on asymptotic expansions to approximate precision measures such as the variance of nonlinear functions of the structural parameters.

3 Optimality vs. Efficiency

Sampling theory approaches lead to efficient estimators (if the proper estimation method is used) conditional on the model specification including any parametric assumptions. A Bayesian approach leads to optimal estimators relative to a specified loss function and any parametric assumptions; uncertainty regarding model specification, nuisance parameters, or even parametric assumptions can be integrated out. Bayesian estimators are referred to as *optimal* rather than efficient because the concept of efficiency does not translate well to the Bayesian paradigm. The Bayesian estimators are optimal in the sense that they combine prior and sample information according to an optimal rule (Bayes' theorem), resulting in a posterior distribution or posterior minimum expected loss estimator that has properly (and optimally) accounted for all available information. This is clearly analogous to the concept of statistical efficiency; however, it is not identical. Because the posterior distribution depends on a prior, which can be different for each researcher, and because point estimators are conditional on the particular loss function specified, a unique minimum variance standard such as the Cramer-Rao lower bound does not exist in an objective sense.

This does not mean that Bayesian estimators are inefficient, only that the way in which their performance is evaluated is slightly different from sampling theory estimators. In fact, I would argue that the Bayesian optimality property is better than the sampling theory's efficiency in the world of applied econometrics because the biggest assumption made is generally that of model specification. Efficiency only holds conditional on model specification, and correctly specifying the model is close to a zero probability event, so the appeal of statistically efficient estimators in applied work is somewhat limited (it seems to me). I would rather know that for the model and prior distribution postulated, right or wrong, the Bayesian estimator (or posterior distribution) makes the best possible use of all information available.

4 Odds Ratios vs. p-Values

Bayesian and sampling theory approaches provide information regarding hypothesis tests in two very different forms. Bayesian procedures produce posterior probabilities supporting each entertained hypothesis (which can be more than two at one time). These posterior probabilities directly present the relative strength of belief in each hypothesis after proper consideration of prior and sample information; that is, the probability that each theory is "true" or a better representation of the underlying process than all other models placed under consideration. When there are only two entertained hypotheses, these posterior probabilities are often presented in the form of a posterior odds ratio (defined in equation 2.8). Again, the posterior odds ratio is directly related to the relative support for each hypothesis, with the value of unity marking the point where the available information fails to distinguish between the two hypotheses. Values greater or less than 1 present the strength of support in favor of one of the hypotheses with the relationship between the posterior odds ratio and the support for the hypotheses being a monotonic function.

Sampling theory hypothesis tests present evidence in the forms of test statistics and p-values based on the assignment of a null hypothesis under which the distribution of the test statistic can be derived. Conditional on the null hypothesis being true, the p-value (associated with the test statistic) describes the probability of observing the given sample of data. It is not a measure of the data's support for the null hypothesis relative to an alternative hypothesis, but simply a probabilistic measure of the empirical discrepancy between certain observed data characteristics and those expected under the null hypothesis.

A simple summary can be made of the two approaches to providing information concerning hypotheses. Sampling theory results state: conditional on the null hypothesis being true, here is the probability of seeing such an empirical discrepancy in a random sample of data. Bayesian results state: conditional on the prior and sample data information, here is the probability support for a particular hypothesis relative to clearly specified alternative hypotheses. Thus, Bayesians measure the data's support for the hypothesis, while sampling theorists measure the hypothesis's support for the data.

5 Hidden or Incomplete Assumptions

While the sampling theory approach does not create hidden assumptions, it also does not force a researcher to be as forthcoming as the Bayesian approach does. When performing Bayesian analysis, one must clearly state all prior information (assumptions), clearly detail the model or models considered, make clear the loss function used to derive any point estimates, and fully describe the likelihood function of the data. A good sampling theory application will also contain all these items, but it is easier to become lazy in the sampling theory approach and ignore paying careful attention to one or more of these details.

Decision Theory

Bayesian methods can also be used in the field of decision science. In this realm we are not interested in trying to estimate some unknown random variable, but in trying to choose the optimal value of some (deterministic) control variable such as the level of capital investment, a quantity of consumption, or an allocation of resources. If all parameters involved in the decision process are known and deterministic, choosing the optimal level of control variables is a simple mathematical exercise. However, in most cases involving economic decisions many of the parameters are unknown random variables; this is where a Bayesian approach becomes useful. By taking account of all available prior and data-based information to construct posterior distributions for the parameters of the optimization problem, levels can be found for the control variables that maximize or minimize the expected value of the objective function (Berger, 1985). It is common in the Bayesian decision theory literature to refer to minimizing the expected loss of a decision or to minimizing the expected risk (Schervish, 1995); however, in the context of economics it is more intuitive to think in the equivalent terms of maximizing the expected value of the decision. The goal of a Bayesian decision theory exercise is to make a decision that is robust with respect to unknown parameter values by selecting the level of the control variables through a process that accounts for all possible (in terms of posterior probability support) values of the random parameter values and weighs these possible parameter values by the probability of their occurrence.

A simple example of a decision theory problem that is well-suited to a Bayesian approach is the case of a firm choosing its productive capacity. Assume that the firm produces a single output using two inputs: one raw material and a capital input that tranforms the raw material into the finished product. The capital input (machinery) lasts forever and can be bought or sold once a year for a constant price. Production is deterministic and occurs daily; however, consumer demand is stochastic and the demand curve is not known with certainty by the firm (although the firm does have data on sales, prices, and other related variables). The choice of the level of capital to own in a particular year is a perfect example of selecting a control variable's level in an environment where the resulting value of that decision is random (due to the random variables present in consumer demand). Using a Bayesian approach to select the productive capacity allows the firm to fully incorporate the uncertainty involved in the random parameters while maximizing the expected value of any objective function that reflects the firm's goals (profit, utility of profits, total sales, etc.).

Let the objective function that the firm wishes to maximize be $U(p, q, w, k)$ where p is the price per unit of the finished product, q is the quantity sold, w is the cost per unit of the raw material, and k is the level of productive capacity, and let the consumer demand schedule facing the individual firm be represented by the function $q = D(p, z|\theta)$ where z is a vector of nonprice variables that influence the level of demand and θ is a vector of unknown parameters of this function. Represent the posterior joint distribution of the random variables in the demand function by $p(\theta|p, q, z, I)$ where I denotes the firm's prior information. The firm's decision problem can then be described as trying to choose the value of k that maximizes

$$\int U(p, D(p, z|\theta), w, k)p(\theta|p, q, z, I)d\theta. \qquad (2.17)$$

The integral in equation (2.17) can be evaluated either analytically or by one of the numerical methods, such as Monte Carlo integration, discussed in depth in Chapter 3. The firm can simply scan over values of k to find the level that maximizes (2.17) or differentiate with respect to k and proceed with the solution of the first-order condition. The explicit incorporation of the posterior distribution of the random variables (which includes prior information) makes this decision theory approach Bayesian.

The classical approach to such a problem has been the "plug-in" method, where the sampling theory estimates of any unknown random parameters are inserted into the optimization problem as if they are the true values and then the problem is solved. The certainty equivalence theorem demonstrates that if the distributions of the random parameters are symmetric, any constraints (such as the demand curve in the preceding example) are linear, and the objective function is linear-quadratic, the plug-in approach yields the same optimal control as a methodology that fully accounts for the stochastics involved. However, prior information or particular distributional assumptions can yield asymmetric posterior distributions for random variables, and in many instances the objective function best suited to a problem is not quadratic (such as a negative exponential utility function for profit). In such cases, taking a Bayesian approach to solving for the optimal control will produce a different answer from the standard method. Bayesian methods to solving such problems are ideal for cases where economic theory provides prior information that shapes posterior distributions (such as demand curves slope down) and for problems where risk is important and the correct objective function is not quadratic, but skewed. All types of risk attitudes can be readily incorporated, and uncertainty concerning underlying parameters is handled in a direct and simplistic manner.

Another advantage of the Bayesian decision theory approach to such problems is that when the optimal solution is found by scanning over a range of control settings, the associated expected values of the objective function can be used to determine a range of near optimal settings of the control variable and to calculate the expected loss attached with suboptimal settings of the control.

References

Berger, J. O. (1985) *Statistical Decision Theory and Bayesian Analysis*, New York: Springer-Verlag.

Box, G. E. P., and G. C. Tiao (1973). *Bayesian Inference in Statistical Inference*. Boston: Addison-Wesley. Now also New York: John Wiley & Sons (1992).

Dorfman J. H., and A. M. Havenner (1992). A Bayesian approach to state space multivariate time series modeling. *Journal of Econometrics 52*, 315–346.

Schervish, M. J. (1995). *Theory of Statistics*. New York: Springer-Verlag.

Zellner, A. (1971). *An Introduction to Bayesian Inference in Econometrics*. New York: John Wiley & Sons.

3

New Advances in Numerical Bayesian Techniques

The numerical techniques discussed in this chapter are designed to free researchers from dependence on analytical methods for the solution of problems concerning probability distributions. Rather than having to use calculus to find the mean, median, mode, or some specified percentile of a distribution, we can now use computer-intensive methods based on (pseudo-)random number generation to empirically investigate the properties of the distributions that arise within Bayesian analyses of estimation and decision problems. Four basic methods for such numerical analysis of probability distributions will be presented in this chapter: Monte Carlo sampling, antithetic replication, importance sampling, and Gibbs sampling.

All of these methods operate on the same basic principle. Using a computer to generate random numbers from a specified distribution, the value of the probability distribution function, along with some number of functions of the random variables involved, is calculated. The value of the pdf and the other functions of interest are saved after each evaluation, and the process repeats itself until a large number of points in the sample space of the random variables have been visited. At this point empirical distributions of the random variables and associated functions have been constructed and permit simple inspection of various properties of these empirical approximations to the true underlying analytical distributions. A researcher can find the mean, median, mode, or a specific percentile (such as the quartiles); compute variances; calculate the probability support for a specified hypothesis; and so on. Anything that could be done analytically if the problems were tractable to such an approach can be done numerically. In addition, while using such a numerical approximation does lead to some level of approximation error in the empirical results, the size of the approximation error is a function of the number of points in the distribution that are evaluated and so is completely within the researcher's control. These concepts and methods will be made clearer through the following discussion of the four specific numerical methods.

The four methods covered in this chapter by no means provide an exhaustive coverage of numerical Bayesian techniques. Numerical approximations can be constructed without random number generation by mathematical approximation of the true functions involved. Additional techniques also exist which do rely on random number generation to approximate the distributions of interest. Interested readers are referred to Tanner (1996) for details an additional methods.

Monte Carlo Sampling

The simplest form of numerical analysis is based on Monte Carlo sampling. The use of these methods in economics can be traced to important works by Hammersley and Handscomb (1964), Kloek and van Dijk (1978), and Geweke (1986, 1991). Monte Carlo sampling relies on direct numerical evaluation of the integrals or probability distributions of interest at a large number of points in the relevant sampling space instead of analytical solution using mathematical equations. The basic process has five steps. To help make the following description concrete, assume that we are analyzing the posterior distribution $p(\theta|y, X)$ where θ is a k-vector random variable, y is an n-vector of observations on the dependent variable, and X is an $(n \times k)$ matrix of (exogenous) explanatory variables.

The Five Steps to Monte Carlo Sampling

1. Draw a random value for the parameter vector θ from the distribution $p(\theta|y, X)$. Denote this computer-generated random draw by $\theta^{(i)}$.
2. Compute the value of any functions $g(\theta^{(i)})$ that are of interest.
3. Save the values of the parameter vector $\theta^{(i)}$ and other functions computed in a matrix that is filled row by row with the values from step 2.
4. Repeat steps 1 to 3 many times, $i = 1, 2, \ldots, B$, where B might be 5000 or 10,000.
5. Using the saved values for $\theta^{(i)}$ and $g(\theta^{(i)})$, compute means, medians, interquartile ranges, and so on, as desired.

Because all draws are made randomly from the correct posterior distribution of θ, they can be treated as equally likely events. Thus, the posterior mean of θ is numerically approximated as the simple arithmetic mean of the B draws compiled in steps 1 to 4; posterior means of functions of the parameters $g(\theta)$ are calculated in an identical manner using their B saved values. The posterior median of either θ or $g(\theta)$ is found by sorting the B saved values from small to large and selecting the middle value as the estimate of the posterior median. If B is an even number, one can use the simple average of the $B/2$ and $(B + 1)/2$ ordered values. Interquartile ranges or other percentiles of the posterior distributions of either θ or $g(\theta)$ are found in an analogous manner to the median. Denoting the numerical estimate of the posterior mean of $g(\theta)$ using B Monte Carlo draws by $\hat{g}_B(\theta)$, Geweke (1989) has shown that under fairly innocuous conditions $\hat{g}_B(\theta)$ converges almost surely to $E[g(\theta)]$ as B tends to infinity. Thus, these numerical approximations are consistent estimators of the corresponding analytical Bayesian estimators, and numerical Bayesian estimators retain the same optimality property as all other Bayesian estimators.

Measures of precision for these numerical estimates can be found in a straightforward manner using the same B saved values for the parameters and their associated functions. The estimated standard error of the numerical approximation in $\hat{g}_B(\theta)$

is given by (Geweke, 1989)

$$s_{na} = \frac{1}{\sqrt{(B)}} \sqrt{\frac{\sum_{i=1}^{B} \left[g(\theta^{(i)}) - \hat{g}_B(\theta)\right]^2}{B-1}}. \tag{3.1}$$

Equation (3.1) makes clear that the size of the error due to numerical approximation can be controlled by the choice of B, the number of random draws generated from $p(\theta|y, X)$.

Monte Carlo sampling is the simplest of the four numerical techniques covered in this book. All posterior estimators for the parameters and any functions of those parameters can be evaluated by simple averages or by sorting the saved draws that form the empirical distribution. The advantage of such an approach is that to find the analytical posterior distribution of a complex function $g(\theta)$ can be quite difficult, involving advanced calculus to perform the change in variables (the sample space through which we are viewing the probability support for a random variable). However, it is generally straightforward to evaluate any function $g(\theta)$ given the condition that $\theta = \theta^{(i)}$. By drawing a sample on θ and using it to construct the posterior distribution of $g(\theta)$, the mathematical difficulties involved in the derivation of this distribution are avoided at the expense of the computer time used to build the distribution empirically.

Antithetic Replication

A slight variation on Monte Carlo sampling that produces an increase in the efficiency of the numerical approximation is the method of antithetic replication. Antithetic replication is not a method for the numerical evaluation of probability distributions or their moments, per se, but is a method that can be added to a standard Monte Carlo sampling scheme. The introduction of antithetic replication adds an additional step to the basic process outlined earlier designed to ensure that the points visited throughout the posterior distribution are evenly distributed around the mean of the distribution. After making a random draw from the posterior distribution, an antithetic replicate is created that is a mirror image of the latest draw, projected through the mean. Making this concrete, the precise steps are listed next.

The Six Steps to Antithetic Replication

1. Draw a random value for the parameter vector θ from the distribution $p(\theta|y, X)$. Denote this computer-generated random draw by $\theta^{(i)}$.
2. Compute the value of any functions $g(\theta^{(i)})$ that are of interest.
3. Save the values of the parameter vector $\theta^{(i)}$ and other functions computed in a matrix that is filled row by row with the values from step 2.
4. Create an antithetic replicate $\theta^{(-i)} = E(\theta) - [\theta^{(i)} - E(\theta)] = 2E(\theta) - \theta^{(i)}$, where $E(\theta)$ is the mean of the posterior distribution $p(\theta|y, X)$, and follow steps 2 and 3 with this draw, too.

5. Repeat steps 1 to 4 many times, $i = 1, 2, \ldots, B$, where B might be 500 or 1000; note that the total number of observations in the empirical distribution will be $2B$.
6. Using the saved values for $\theta^{(i)}$ and $g(\theta^{(i)})$, compute means, medians, interquartile ranges, and so on as desired.

Calculation of means, medians, percentiles, and measures of precision are exactly as in the case of standard Monte Carlo integration. Almost sure convergence of the numerical estimator to its analytical counterpart still holds. The benefit of antithetic replication is that in many cases an equally precise numerical approximation can be achieved on a much smaller number of observations in the empirical distribution. Equation (3.1) showed that the variance of the numerical approximation error is equal to $(1/B)\sigma^2(g)$ where $\sigma^2(g)$ is the variance of $g(\theta)$, assuming it exists from the posterior distribution. If the sample size of the data used in a particular application is of size T, using antithetic replication can shrink the size of the numerical approximation by a factor of T (Geweke, 1988). That is, the variance of the numerical approximation error under antithetic replication is equal to $(1/BT)\sigma^2(g)$ or $(1/T)s_{na}^2$, where s_{na} is as defined in equation (3.1).

This decrease in the variance of the numerical approximation (due to numerical approximations that converge to the true value at a faster rate) can be quite an advantage in computational efficiency if the sample size is moderately large. In standard Monte Carlo sampling, $B = 5000$ and 10,000 are common choices to ensure suitable numerical accuracy. However, if the data set has a sample size of $T = 100$, between 500 and 1000 antithetic random draws (250 to 500 pairs) will generally achieve the same degree of numerical accuracy. The precise increase in numerical accuracy from antithetic replication depends on the amount of nonlinearity in the function $g(\theta)$, with the gains declining as the nonlinearity increases. As an example of the potential gains in sampling efficiency, Geweke (1988) found that with a sample size of $T = 60$ a three-step ahead predictive mean could be estimated with equal numerical accuracy using either 10,000 standard Monte Carlo draws or 550 antithetic draws.

The only additional burden is that the mean of the posterior distribution must be known (generally the mean of θ can be found easily; it is the properties of some function $g(\theta)$ that forces the reliance on numerical rather than analytical methods). If the mean of θ is not known, it can be estimated by a preliminary Monte Carlo sampling routine that does not bother to calculate additional functions of the parameters. This might make sense if the functions of θ to be evaluated are complex and time-consuming and there is a large time-savings to be realized by the reduction in the number of draws necessary when using antithetic replication.

Importance Sampling

The third numerical method to be covered is importance sampling. This is the most useful method for analyzing common regression models with Gaussian likelihood functions but informative priors that are not conjugate; thus, the posterior distribution is of a nonstandard form from which it is difficult or impossible to generate random draws. Importance sampling allows the random draws of θ to be generated from a substitute density and the empirical distribution is then adjusted to account for the differences between the substitute density and the actual posterior distribution of θ. The steps to an importance sampling algorithm are as follows.

The Six Steps of Importance Sampling

1. Draw a random value for the parameter vector θ from a substitute density $f(\theta)$. Denote this computer-generated random draw by $\theta^{(i)}$.
2. Compute the values of $f(\theta^{(i)})$ and $p(\theta^{(i)}|y, X)$.
3. Compute the value of any functions $g(\theta^{(i)})$ that are of interest.
4. Save the values of $\theta^{(i)}$, the substitute density, the posterior distribution, and other functions computed in a matrix that is filled row by row with the values from steps 1 to 3.
5. Repeat steps 1 to 4 many times, $i = 1, 2, \ldots, B$, where B might be 5000 or 10,000.
6. Using the saved values for $\theta^{(i)}$ and $g(\theta^{(i)})$, compute means, medians, interquartile ranges, and so on as desired, using the importance weights to correct for the differences between the substitute and true posterior distribution of θ.

Generation of the random draws from a substitute density $f(\theta)$ results in an empirical sample of observations on θ that are not all equally likely under $p(\theta^{(i)}|y, X)$; i.e., it is not a random sample from the posterior distribution. This means that simple averages cannot be used to estimate the posterior mean of either θ or $g(\theta)$; weighted averages must be used instead to correct for the nonrandom sample. The correction is simple and uses the values of the two densities at each point in the empirical distribution to form the weights. These weights, often called importance weights, are given by the ratio $p(\theta^{(i)}|y, X)/f(\theta)$. To make this clear, the formula for the posterior mean of a function $g(\theta)$ using importance sampling is

$$\hat{g}_B^{IS}(\theta) = \frac{\sum_{i=1}^{B} g(\theta^{(i)})p(\theta^{(i)}|y, X)/f(\theta^{(i)})}{\sum_{i=1}^{B} p(\theta^{(i)}|y, X)/f(\theta^{(i)})} = \frac{\sum_{i=1}^{B} g(\theta^{(i)})w(\theta^{(i)})}{\sum_{i=1}^{B} w(\theta^{(i)})}, \quad (3.2)$$

where $w(\theta^{(i)})$ is the importance weight for the ith observation in the empirical distribution and the superscripted IS denotes that this estimator is based on importance sampling rather than on direct Monte Carlo sampling. Calculation of percentiles of the posterior distribution of $g(\theta)$, such as the median and interquartile range, can be done by sorting, from smallest to largest, the B observations on $g(\theta)$ with their associated values of $w(\theta^{(i)})$. Then one simply adds the individual values of $w(\theta^{(i)})$ until the sum reaches the desired percentile (e.g., 0.50 for the median); the corresponding saved value of $g(\theta^{(i)})$ is the estimate of that percentile. Just remember to normalize by the sum of the importance weights $w(\theta^{(i)})$. Again, the numerical approximation to $g(\theta)$ converges almost surely to the expected value of $g(\theta)$ as B tends to infinity given certain conditions on the choice of the substitute density (Geweke, 1989).

The choice of the substitute density is important in terms of both accuracy of the numerical approximation and computational efficiency (how many draws are needed to achieve an estimate of the desired accuracy level). The more similar the two densities are, the better the numerical approximation will be. The most important point in the selection of a substitute density is to ensure that the tail coverage of the substitute density is at least equal to, if not slightly greater than, the tail coverage of the true posterior distribution. That is, the substitute density should have somewhat fatter tails than the posterior density of θ. This leads to oversampling in the tails of the posterior distribution, but these observations are downweighted by the importance weight function $w(\theta)$. If the tails are too thin, you will get inadequate coverage in the tails, and upweighting of these observations may not be enough to compensate for the failure to fully visit the entire range of the sample space in the numerical sampling scheme (Gewke, 1989).

It is therefore common to use substitute densities with scaled-up variance-covariance matrices to ensure proper coverage (Tanner, 1996, p. 57). For example, consider a single equation regression model with n observations and k exogenous variables, $y = X\beta + \varepsilon, \varepsilon \sim N(0, \sigma^2 I)$ and the prior distribution for $\theta = (\beta, \sigma^2)$ is proportional to $h(\beta)/\sigma$. Thus, we have an informative prior on the regression coefficients β and a standard diffuse prior on the scale parameter. The likelihood function of the data is of a standard multivariate normal form and the posterior distribution would also have a standard form (multivariate normal-inverse gamma for $\theta = (\beta, \sigma^2)$) if $h(\beta)$ were a constant (Zellner, 1971, p. 67). However, in many economic cases $h(\beta)$ is far from being a constant across the sample space of β; it may be truncated by economic theory, or it may simply be informative due to knowledge gained from previous research. In many such cases, importance sampling is a natural candidate to use in the examination of the posterior distribution of β or $g(\beta)$. The likelihood function of the data is often a natural choice for the substitute density, assuming that the form of the prior distribution does not cause the likelihood and the posterior to be radically different (in the region of the sample space that receives prior support). Other common choices are the multivariate normal and Student-t distributions. Random draws on (β, σ) can be generated easily, then the prior distribution is evaluated, and the entire set of steps listed earlier is followed to complete the analysis. With reference to the preceding remarks concerning tail coverage, past research studies have often used a scaled value of the maximum likelihood estimate of σ (or Σ for cases with non-iid errors) to ensure adequate sampling in the tails of the posterior distribution. Tanner (1996, p. 57) and the author have both had success with setting the variance-covariance matrix to 1.5 times that of the maximum likelihood estimate when using a data-centered normal distribution to generate the empirical sample of draws on β.

The numerical approximation error can be estimated in a manner similar to that used with Monte Carlo sampling. Denoting the importance sampling estimate of the posterior mean of $g(\theta)$ by $\hat{g}_B^{IS}(\theta)$, the estimated standard error of the numerical approximation in $\hat{g}_B^{IS}(\theta)$ is (Geweke, 1989):

$$s_{na} = \frac{\sqrt{\sum_{i=1}^{B}\left\{w(\theta^{(i)})\left[g(\theta^{(i)}) - \hat{g}_B^{IS}(\theta)\right]\right\}^2}}{\sum_{i=1}^{B} w(\theta^{(i)})} \tag{3.3}$$

Equation (3.3) makes clear that the size of the error due to numerical approximation will be inflated as the values of $w(\theta^{(i)})$ vary from unity (because quadratic functions are convex). Thus, choosing a substitute density that is a good approximation of the actual posterior distribution $p(\theta|y, X)$ will produce a better estimate of the posterior means, medians, and so on of both θ and its functions.

It is worth noting that unlike with standard Monte Carlo sampling, using a large number of draws to numerically evaluate the posterior distribution does not guarantee a small numerical approximation error. The formula in equation (3.3) for s_{na} is not a function of B in any direct sense, let alone a decreasing function of B. If all the values of $w(\theta^{(i)})$ equal 1, the expression in equation (3.3) converges to that in equation (3.1) as B goes to infinity (the only reason they are not identical for finite B is the presence of $(B-1)$ instead of B in the denominator of equation (3.1). Intuition therefore suggests that if the $w(\theta)$ function remains close to 1 for most draws, the numerical accuracy will increase with the number of draws B, but researchers should be cautioned that to get such benefits from large samples of observations on the posterior distribution they must use a substitute density that provides good coverage relative to the true posterior distribution.

Finally, antithetic replication can be used in conjunction with importance sampling, too. Simply combine the steps for importance sampling with those outlined for antithetic replication. This will tend to increase the numerical efficiency for a given number of observations on the posterior distribution for any symmetrical or near-symmetrical posterior distribution (it does not matter whether the substitute density is symmetrical). An important step in any importance sampling with any antithetic replication algorithm is to make sure that the value of $p(\theta^{(-i)}|y, X)$ is calculated for the antithetic replicate. By definition, $f(\theta)$ is symmetric or we would not be using antithetic replication, so $f(\theta^{(i)}) = f(\theta^{(-i)})$; however, $p(\theta|y, X)$ is likely to be asymmetric (remember that we are resorting to importance sampling due to some type of complexity in $p(\theta|y, X)$) and thus the probabilities of these paired parameter vectors will generally be different.

Gibbs Sampling

A somewhat different approach to the numerical methods presented so far for generating random draws from a probability distribution is developed by a class of methods referred to as *Markov chain Monte Carlo algorithms*. The inclusion of the words *Markov chain* indicates that instead of generating a large sample of independent draws from the posterior distribution (Monte Carlo sampling) or substitute density (importance sampling), each draw is related to (conditioned on) the previous one. At first glance, it may seem that generating a sequence of correlated draws will lead to difficulties in accurately assessing expectations of parameters or functions of the parameters; however, it can be shown that when done properly the chain of correlated draws can, in fact, be used to consistently estimate any desired expected value (or other estimators such as distribution percentiles). The most commonly used Markov chain Monte Carlo algorithm is the Gibbs sampler.

Gibbs sampling was first proposed by Geman and Geman (1984). The basic idea behind Gibbs sampling is to generate draws from a multivariate joint probability distribution by sequential sampling from a series of conditional probability distributions on lower-dimensional subsets of the random variables. In cases where it is difficult or impossible to generate random draws from the joint posterior distribution $p(\theta|y, X)$, Gibbs sampling takes advantage of the fact that it is often possible to generate draws from a set of conditional probability distributions on these lower-dimensional subsets of random variables, due to simplifications that occur in the forms of the distributions. Consider a posterior distribution $p(\theta|y, X)$ where θ can be partitioned into three lower-dimensional parameter vectors, θ_i, $i = 1, 2, 3$. Using the partitioned θ parameter vector as an example, such a Gibbs sampling algorithm would consist of the following steps.

The Five Steps to Gibbs Sampling

1. Begin with some initial values (guesses) for the parameter vectors θ_1, θ_2, θ_3; denote these initial values by $\theta_i^{(0)}$, $i = 1, 2, 3$.
2. Generate random draws in sequence from the conditional posterior distributions:

$$\theta_1^{(j+1)} \sim p(\theta_1|y, X, \theta_2^{(j)}, \theta_3^{(j)})$$
$$\theta_2^{(j+1)} \sim p(\theta_2|y, X, \theta_1^{(j+1)}, \theta_3^{(j)})$$
$$\theta_3^{(j+1)} \sim p(\theta_3|y, X, \theta_1^{(j+1)}, \theta_2^{(j+1)})$$

3. Repeat step 2 many times, conditioning at each iteration on the most recently generated parameter vectors for the other partitions. Discard the first J triplets of $(\theta_1^{(j)}, \theta_2^{(j)}, \theta_3^{(j)})$ to avoid dependence on the initial values. Then save the next B triplets.
4. Check the chain for convergence to ensure it is safe to stop generating empirical observations on θ (details given later).
5. Using the saved values for $\theta^{(j)}$, compute means, medians, and interquartile ranges of θ or any desired functions $g(\theta)$. The saved values for each full iteration on step 2 can be treated as independent, identically distributed random draws from the full joint posterior distribution of θ for the purposes of computed

posterior means; that is, you can use the same simple arithmetic formulas as for standard Monte Carlo sampling without any sort of correction as is needed with importance sampling.

One simple, effective method for checking convergence of the Gibbs sampler is based on the ratio of two estimated densities (Tanner, 1996, p. 157). Let $q(\theta_1^{(j)}, \theta_2^{(j)}, \theta_3^{(j)}|y)$ be a function that is proportional to the current approximation to the conditional joint posterior density, $p(\theta_1^{(j)}, \theta_2^{(j)}, \theta_3^{(j)}|y)$ evaluated using the ith Gibbs draw; that is, the normalizing constant is not necessary. Let $f(\theta_1^{(j)}, \theta_2^{(j)}, \theta_3^{(j)})$ be the current approximation to the unconditional joint posterior density, $p(\theta_1^{(j)}, \theta_2^{(j)}, \theta_3^{(j)}) = p(\theta_1^{(j)}, \theta_2^{(j)}, \theta_3^{(j)}|y, X)/p(y)$. Then denote the Gibbs stopping function $m(j) = q(\theta_1^{(j)}, \theta_2^{(j)}, \theta_3^{(j)}|y)/f(\theta_1^{(j)}, \theta_2^{(j)}, \theta_3^{(j)})$ When the Gibbs sampler converges, the function $m(j)$ will converge toward a degenerate distribution with value equal to the $q()$ function's missing normalizing constant. Thus, the stopping function can be evaluated after each completion of a full iteration (or after every k iterations) and checked for convergence. While the number of iterations to achieve satisfactory convergence with a Gibbs sampler is much more variable and harder to predict than with Monte Carlo sampling, it is reasonable with modern computer speeds to plan on discarding the first 500 or 1000 iterations and then to evaluate the posterior based on the next 2000 to 5000 iterations of the Gibbs sampler. Good references that discuss the theory or methods of testing convergence of the draws to the underlying joint distribution include Chan (1993), Chib (1995), Geyer (1992), Schervish and Carlin (1992) and Tierney (1994).

Gibbs sampling can also be very useful for applications of data augmentation. Data augmentation is used when latent variables are treated as random parameters and observations for them, generated from their probability distributions, are treated as data to be used in the empirical investigation. Applications using dichotomous dependent variable models, state space, or common factor models, and models containing certain expectational variables are naturals for this treatment. The latent variables often have a conditional distribution that is straightforward to draw from given values for the random parameters of the model. Thus, given a suitable partitioning of the parameter vector into subsets with conditional distributions that are of standard forms, the latent variables can be treated as a further partition of the whole set of random variables whose full joint probability distribution we wish to investigate. In this way, the Gibbs sampler can generate the marginal posterior distribution of the latent variables along with the posterior distribution of the random parameters. Inferences can then be made with regard to any of the random variables without having to encounter any of the difficulties that may arise from the treatment of latent variables as some sort of special class.

Dorfman (1996) and McCulloch and Rossi (1994) are econometric application papers that make concrete the application of the Gibbs sampler to the problem of estimating the parameters of a multinomial probit model (with its inherent latent variables) and examining the posterior distributions of functions of the model parameters (such as the marginal probabilities of a particular choice).

References

Chan, K. S. (1993). Asymptotic behavior of the Gibbs sampler. *Journal of the American Statistical Association* 88, 320–326.

Chib, S. (1995). Marginal likelihood from the Gibbs output. *Journal of the American Statistical Association* 90, 1313–1321.

Dorfman, J. H. (1996). Modeling multiple adoption decisions in a joint framework. *American Journal of Agricultural Economics* 78, 547–557.

Geman, S., and D. Geman (1984). Stochastic relaxation, Gibbs distributions, and Bayesian restoration of images. *IEEE Trans. Pattern Anal. and Machine Intelligence* 6, 721–741.

Geweke, J. (1986). "Exact inference in the inequality constrained normal linear regression model." *Journal of Applied Econometrics* 1, 127–141.

Geweke, J. (1988). Antithetic acceleration of Monte Carlo integration in Bayesian inference. *Journal of Econometrics* 38, 73–89.

Geweke, J. (1989). Bayesian inference in econometric models using Monte Carlo integration. *Econometrica* 57, 1317–1339.

Geweke, J. (1991). "Generic, algorithmic approaches to Monte Carlo integration in Bayesian inference." *Contempory Mathematics* vol.115, eds. N. Flournoy and R. K. Tsutakawa. Providence: American Mathematical Society.

Geyer, C. (1992). Practical Markov chain Monte Carlo. *Statistical Science* 7, 473–482.

Hammersley, J. M., and D. C. Handscomb (1964). *Monte Carlo Methods.* London: Chapman and Hall.

Kloek, T., and H. K. van Dijk (1978). Bayesian estimates of simultaneous equation system parameters: An application of integration by Monte Carlo. *Econometrica* 46, 1–19.

McCulloch, R., and P. E. Rossi (1994). An exact likelihood analysis of the multinomial probit model. *J. Econometrics* 64, 207–240.

Schervish, M. J., and B. P. Carlin (1992). On the convergence of successive substitution sampling. *Journal of Computational and Graphical Statistics* 1, 111–127.

Tanner, M. A. (1996). *Tools for Statistical Inference: Methods for the Exploration of Posterior Distributions and Likelihood Functions, third edition.* New York: Springer.

Tierney, L. (1994). Markov chains for exploring posterior distribution (with discussion). *Annals of Statistics* 22, 1701–1762.

Zellner, A. (1971). *An Introduction to Bayesian Inference in Econometrics.* New York: John Wiley & Sons.

Part II

Applications in Econometrics

4

Imposing Economic Theory

In many applications, economic theory provides information that restricts the sample space that should receive prior support. While the prior distributions implied by economic theory are not always informative within a specific range (subspace), they tend to have truncations that are caused by information contained in the economic theories that are the basis of much of our applied work. The simplest example is economists' general belief that demand curves slope down (at a lower price, the quantity sold should be larger). If this is taken to be a certain feature of economic behavior, then the prior distribution on the parameter vector of a demand regression model should be restricted to support only the region of the parameter space that is consistent with such behavior.

At times, economic theory provides more complex prior information, such as adding up constraints, convexity conditions, or concavity conditions to be met, which are still best represented by truncations on the space receiving support from the prior distribution. In other applications, applied economic knowledge may provide even more information with regard to the unknown random parameters and more informative priors can be used. An example would be a regression model of demand for a commodity, where the commodity is well-established to have an inelastic demand. Economic theory suggests that the price elasticity of demand should be negative, and in this instance we also have knowledge suggesting that values between 0 and -1 are more likely than those with greater absolute values. Estimation of models in situations such as the ones mentioned earlier are well-suited to the application of Bayesian analysis because the prior information can be optimally combined with the information contained in the data to yield a posterior distribution that is consistent with emprirical observations and existing economic theory. Further, it is straightforward to compute the observed support for any restrictions from economic theory that are imposed in the form of prior information, thus providing a check on both the state of economic theory and the agreement between the researcher's prior distribution and the current data set.

This chapter will present a variety of applications in which a Bayesian approach allows easy and productive incorporation of prior information derived from economic theory into econometric estimation of an economic model. The applications will range from extremely simple to fairly complex. Some applications will be described in full detail to demonstrate precisely how a researcher would use such an approach, while others will be mentioned simply as candidates for this method of analysis.

Restricted Parameter Spaces

The simplest type of prior information to incorporate via the prior distribution is a restriction on the parameter space due to a theory that places inequality constraints on the value of a parameter or function of parameters. This type of prior information implies that one should start with some type of standard prior distribution and then multiply that by an indicator $(0/1)$ function that limits positive prior support to the region that satisfies the inequality restriction. This use of Bayesian analysis in applied econometrics is well-demonstrated by Chalfant (1993), Chalfant, Gray, and White (1991), Geweke (1986), and Hayes, Wahl, and Williams (1990).

To begin, consider a single equation demand model of the form

$$q = \beta_1 + p\beta_2 + m\beta_3 + r\beta_4 + \varepsilon = X\beta + \varepsilon, \tag{4.1}$$

where q is a $(T \times 1)$ vector of the quantity sold in each of T time periods of some good; p, m, and r are conformable vectors of the corresponding values for price of the good, disposable income of the product's consumers, and the price of a substitute good; and ε is a vector of stochastic error (discrepancy) terms. The four-element β vector after the second equality sign contains the four individual random parameters, and X is the matrix of four regressors (including the intercept). Economic theory suggests that the signs of the four elements in the β vector should be $(+, -, +, +)$ where the fourth sign is predicated on the definition of the last regressor as the price of a substitute good (not a complement), and the sign of the intercept could depend on the values of the nonprice variables but will be assumed positive here.

Imagine that the elasticity of this product either is not well known in advance or is hard to pin down due to variation in price and quantity (because this is not a constant-elasticity model). In fact, a very weakly informative, but proper prior is desired that incorporates the sign restrictions derived from economic theory. Because little is known about the parameter values, it seems reasonable to assume that the prior distributions of the individual β_i are independent; that is, the marginal prior distribution of each is equal to its conditional prior. This is often a good assumption in applied econometrics. If data is scaled so that all four of the parameters are likely to be of a small order of magnitude (0–10), a suitable prior might be

$$p(\beta, \sigma) \propto \Phi(\beta_1/10)\Phi(\beta_2/10)\Phi(\beta_3/10)\Phi(\beta_4/10)D(\beta_1, \beta_2, \beta_3, \beta_4)/\sigma, \tag{4.2}$$

where σ is the standard deviation of the error term, $\Phi()$ is the probability density of the standard normal distribution, and $D(\beta_1, \beta_2, \beta_3, \beta_4)$ is an indicator function that equals 1 only when all four parameters satisfy the given sign restrictions and equals 0 otherwise. Scaling the regression parameters within the prior distribution implies that the standard deviation of our prior with respect to each of these four parameters is 10. Thus, the prior variances of the β_i are quite large in conformance with our expressed ideal of being only weakly informative. Figure 2 shows for a single β_i parameter what this prior distribution looks like for priors with standard deviations of 1, 5, 10, and 100. Finally, note that the prior on σ takes the standard Jeffreys form for a parameter ranging from 0 to infinity (i.e., it is uninformative with respect to the natural log of σ).

Having fully specified the prior on $\theta = (\beta, \sigma)$, the next step is to specify the likelihood function for the data. If we assume that the error terms are i.i.d. normally distributed, the likelihood takes the standard form of

$$p(y|\theta, X) = (2\pi\sigma^2)^{-T/2} \exp[-0.5(y - X\beta)'(y - X\beta)/\sigma^2]. \qquad (4.3)$$

Multiplying the likelihood function by the prior distribution on θ yields a distribution that is proportional to the posterior distribution of θ:

$$p(\theta|y, X) \propto (\sigma^2)^{-(T+1)/2}\Phi(\beta_1/10)\Phi(\beta_2/10)\Phi(\beta_3/10)\Phi(\beta_4/10)D(\beta)$$
$$\times \exp[-0.5(y - X\beta)'(y - X\beta)/\sigma^2]. \qquad (4.4)$$

Due to the indicator function $D(\beta)$ and the presence of the normal densities in the individual β_i, this posterior distribution cannot be easily dealt with in an analytical framework, nor will Monte Carlo sampling be a useful approach. However, as long as the prior distribution for β is kept fairly diffuse, the posterior distribution is not

Prior distributions for b_i

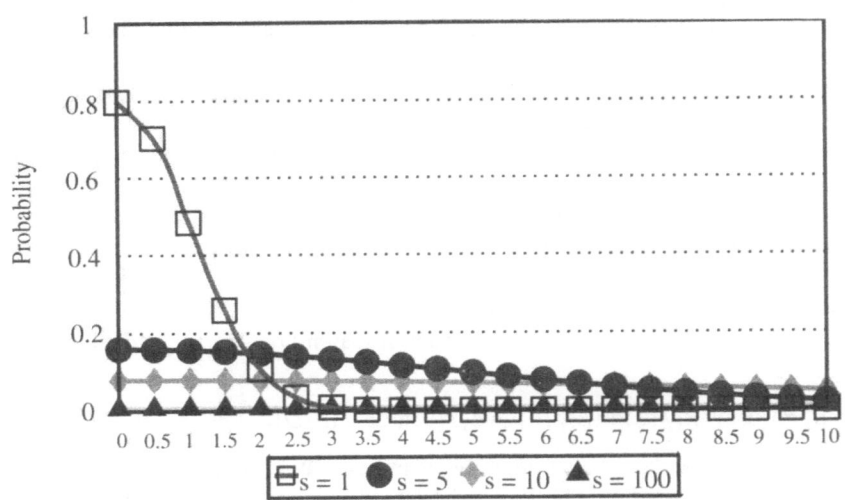

FIGURE 2.

very different from the likelihood function (or a close relative) and importance sampling can be used.

If the posterior distribution of σ^2 is not of particular interest, one might use a multivariate Student-t distribution centered at the mode of the likelihood function (the maximum likelihood estimator of β) as the substitute density; this would be the marginal posterior distribution of β if we had a constant (diffuse) prior on β. If you wish to use the likelihood multiplied by the prior distribution for σ^2 as the substitute density (i.e., the posterior except for the prior on β), a two-step process is used. First, generate a draw of σ^2 from an inverted χ^2 distribution (scaled by $(n - k)s^2$) and then conditional on that draw, generate a value for β from the multivariate normal distribution centered at the MLE again (Geweke, 1988). In the following description, I will assume that the first method is used, generating random draws for β only as that is generally the center of attention in applied work and the example is simpler (which is advantageous for the first one). Choosing this approach, the application would proceed through the following steps.

Application of Importance Sampling to a Truncated Prior Distribution

1. Specify the prior distribution for β and σ. The prior for this example is assumed to be as shown in equation (4.2). Ignore the normalizing constant, as that will be accounted for in the formula for computing the posterior means.
2. Use the data set to estimate β and σ by maximum likelihood, and use these estimates, β_{ML} and σ_{ML}, to specify the substitute density. Let the substitute density be a multivariate Student-t distribution with four (or another small number of) degrees of freedom, mean β_{ML}, and variance-covariance matrix equal to $[T/(T - k)]\sigma_{ML}(X'X)^{-1}$.
3. For $i = 1, 2, \ldots, 10,000$:
   a. Draw a random value for $\beta^{(i)}$ from the substitute density of step 2.
   b. Compute and save $f(\beta^{(i)})$, $p(\beta^{(i)}|y, X, \sigma_{ML})$, and all functions $g(\beta^{(i)})$ of interest.
4. Compute posterior means and medians for $\beta^{(i)}$ and $g(\beta^{(i)})$ using the formula in equation (3.2) for the means. Medians and other percentiles will be discussed more later.
5. Compute the numerical precision of the estimates using the formula for the standard error of the numerical approximation error shown in equation (3.3). This should be presented with the results almost like the common t-values against a zero null to demonstrate the precision of the numerical methods used.
6. Test the hypothesis implied by the restriction on the parameter space derived from economic theory.

To carry out the last step, you need to compute the value of the prior distribution at each point with and without the trucation based on the indicator function $D(\beta)$. Denote the prior on β without the indicator function by $\tilde{p}(\beta)$; $p(\beta) = \tilde{p}(\beta) D(\beta)$. The marginal posterior distribution under this nonrestricted prior would be denoted by $\tilde{p}(\beta|y, X)$. Then the posterior support for the hypothesis embedded in $D(\beta)$ is given by

$$
p(H_1) = \frac{\sum_{i=1}^{10,000} \dfrac{p(\beta^{(i)}|y, X)}{f(\beta^{(i)})}}{\sum_{i=1}^{10,000} \dfrac{\tilde{p}(\beta^{(i)}|y, X)}{f(\beta^{(i)})}}
\tag{4.5}
$$

Equation (4.5) calculates the posterior probability of the restrictions being true conditional on the unrestricted prior specification by counting the number of random draws that satisfy the restrictions. Because importance sampling is used, a simple percentage of draws that satisfy the restrictions cannot be used, so each draw is weighted by the importance weights in a manner exactly like that used in computing the posterior means. The hypothesis can be maintained or rejected on the basis of the value of $p(H_1)$ compared to the posterior support for the alternative hypothesis, which here will equal $1 - p(H_1)$. Since a high expected loss might reasonably be attached to rejecting a well-respected economic theory, one might choose to maintain such a hypothesis unless its posterior support fell well below 0.50, perhaps as low as 0.20.

Note that when working with a truncated prior such as the one postulated here, some of the random draws will be associated with zero values for the posterior distribution $p(\beta^{(i)}|y, X)$. This does not cause any changes in the formulas as they are applied. For example, the posterior mean of the price parameter, β_2, is given by

$$
\bar{\beta}_2 = \frac{\sum_{i=1}^{10,000} \dfrac{\beta_2^{(i)} p(\beta^{(i)}|y, X)}{f(\beta^{(i)})}}{\sum_{i=1}^{10,000} \dfrac{p(\beta^{(i)}|y, X)}{f(\beta^{(i)})}}
\tag{4.6}
$$

This formula will automatically place zero weights on all draws that were not supported by the prior distribution because $p(\beta^{(i)}|y, X)$ will equal 0 for those draws. Posterior means of other functions of the β parameters can be found by similar calculations as long as the value of the function for each of the draws is saved (or calculated from saved parameters as equation (4.6) is applied). For this example, you could calculate the posterior mean of the income elasticity of demand for a particular time period after computing the posterior mean of β_3.

To find the posterior median for a parameter (the optimal point estimate under absolute loss), you must extract two columns from your matrix of saved values from the importance sampling: one column containing the 10,000 parameter values (or function thereof) and the second column containing the importance weights, $w^{(i)} = p(\beta^{(i)}|y, X)/f(\beta^{(i)})$. Then sort these two columns, making sure that the rows stay together, from small to large values for the parameter (function). Next, create a third column equal to the running sum of the column of importance weights; note that many software languages will do this with a preprogrammed command. The values in the column of running sums for the importance weights are the numerical approximations of the percentiles of the distribution of the random variable on which the sort was performed. The median is the parameter value found in the row where the running sum first exceeds 50% (or the last row where it does not). Quartiles and highest posterior density regions can also be found by selecting the proper rows of this matrix of sorted empirical draws. To find the median for a second parameter, you must start from the beginning by extracting a new matrix of that parameter's B randomly generated values and associated importance weights and re-sorting; that is, the sort must be done for each random variable of interest and the importance weights must be handled carefully to ensure that they stay linked with their associated random draws.

To caclulate the numerical approximation errors of the posterior estimates, use equation (3.3), denoting a posterior mean by placing a bar above the random variable's symbol. To compute the numerical approximation error for the posterior mean of the price elasticity for the kth period in the data sample, the formula would be

$$
s_{na} = \frac{\sqrt{\sum_{i=1}^{10,000} \left[w^{(i)} \left(\frac{\beta_2^{(i)} p_k}{q_k} - \frac{\bar{\beta}_2 p_k}{q_k} \right) \right]^2}}{\sum_{i=1}^{10,000} w^{(i)}}
\tag{4.7}
$$

This provides a measure of the precision of the importance sampling algorithm. If the value of s_{na} is suitably small, perhaps 1% of the posterior mean of the elasticity, you can conclude that the numerical integration was successful and that you had enough random draws and a good substitute density.

A final step worth performing in all Bayesian analyses is sensitivity analysis on the prior distribution. In the example just discussed, I might recalculate all estimators under priors with no truncation due to economic theory and with standard deviations in the normal densities of 5, 20, and 100. These results would be calculated using the same set of 10,000 random draws from the importance sampling algorithm. In fact, these other priors can be placed in the computer program at the start so that the implied different posterior density values for each draw are calculated and saved with those of the base prior. Note that the importance weights will be different under the different priors, so a new set of importance weights must be generated for each prior. These other prior distributions are not different possibilities that the researcher considers as alternative candidates; a single person can only have one prior distribution over a particular parameter space. Rather, they allow readers of your research to envision how the results might change under their priors and to satisfy themselves that your results are not completely dependent on the prior you selected. Performing a good sensitivity analysis on your prior distribution convinces readers that your empirical results are robust.

An example of what such sensitivity analysis might look like is shown in Figure 3. With a marginal likelihood for β_3 centered at 2 with a standard deviation of 1

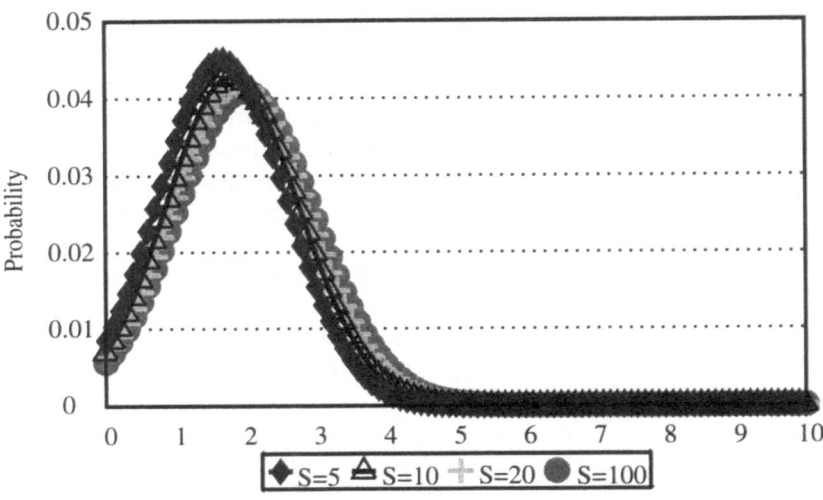

Sensitivity of the posterior distribution of b_3
to different prior variances.

FIGURE 3.

(meaning that a maximum likelihood estimation would yield a t-statistic of 2.0
for the standard zero null hypothesis on β_3), the marginal posterior distribution
of β_3 under four different priors with standard deviations of 5, 10, 20, and 100
are drawn in Figure 3. Note that while the posterior mode (and mean and median)
does change, the effect of the prior variance is clearly slight, and these results can
be declared robust with respect to that aspect of the prior distribution.

A Profit Function Application

Another natural application of numerical Bayesian methods presents itself in the
estimation of systems of equations that arise in the context of profit and cost
function estimation. When estimating a profit or cost function, applied econometricians often estimate the profit or cost function along with a system of input
demand and/or output supply equations. Economic theory enters through two
routes, cross-equation equality restrictions that effectively reduced the number
of random parameters and inequality restrictions on functions of the parameters
that reflect economic theory. A common application in the applied literature has
been estimation of the translog profit function in conjunction with the netput share
equations (either output supply or input demand), and many researchers have imposed convexity on this system following the theory and conditions first stated by

Lau (1976, 1978). Just a few of the applications can be found in Ball (1988), Lopez (1984), and Shumway (1983).

Denoting variable inputs and outputs by y_j and their prices by p_j, $j = 1, J$; fixed inputs by x_k, $k = 1, K$, profit by π; and leaving out the time index that is often included in these models to simplify the presentation here, the translog profit function with fixed inputs can be written as

$$\ln(\pi_t) = \alpha_0 + \sum_{j=1}^{J} \alpha_j \ln(p_{jt}) + \sum_{k=1}^{K} \beta_k \ln(x_{kt}) + \frac{1}{2} \sum_{j=1}^{J} \sum_{s=1}^{J} \gamma_{js} \ln(p_{jt}) \ln(p_{st})$$

$$+ \frac{1}{2} \sum_{j=1}^{J} \sum_{k=1}^{K} \theta_{jk} \ln(p_{jt}) \ln(x_{kt}) + \frac{1}{2} \sum_{s=1}^{K} \sum_{k=1}^{K} \delta_{sk} \ln(x_{st}) \ln(x_{kt}) + v_t, \tag{4.8}$$

where the subscript t denotes the time period of each observation, $t = 1, T$ and v_t is a stochastic error term due to the approximation of the true underlying technology. The corresponding netput demand/supply equations (with a similar additive error e_t added to each one) take the form

$$S_{jt} = \alpha_j + \sum_{s=1}^{J} \theta_{sj} \ln(p_{st}) + \sum_{k=1}^{K} \delta_{jk} \ln(x_{kt}) + e_{jt}, \tag{4.9}$$

where S_{jt} is the gross revenue or cost share of the jth netput, $p_t y_t / \pi_t$. Economic theory suggests that profit functions should be homogeneous of degree 1 which leads to restrictions on the sums of many of the parameters; the symmetry requirement for the matrix of second derivatives provides a set of equality restrictions for more parameters. Incorporating these two pieces of information from economic theory allows the econometrician to drop the Jth netput share equation from the set of equations to be estimated and then to recover all of its parameters from those that are estimated directly. It also reduces the number of random variables to a more manageable number. These restrictions do not affect a Bayesian analysis differently in any way from the sampling theory approach (or the mathematical programming approach, also used on these problems). However, the convexity condition on the profit function is relevant information concerning the random parameters that should be included in the prior distribution constructed for them.

For complicated models such as the one postulated in this application, it is unlikely that the researcher will have prior information concerning the values of the random parameters; thus, a very diffuse informative prior centered at a zero-vector is often appropriate. (In the next chapter, the formation of informative priors on functions of the structural parameters about which we may be better informed will be discussed). As in the preceding example, economic theory suggests that we modify this prior by an indicator function that restricts prior support to the region where the profit function is convex. The matrix of second derivatives of profit with respect to the netput prices follows the pattern:

$$\partial^2 \pi_t / \partial p_{it} \partial p_{jt} = (\pi / p_{it} p_{jt})[\gamma_{ij} + S_{it} S_{jt}], \, i \neq j, \quad \text{and} \tag{4.10}$$

$$\partial^2 \pi_t / \partial p_{it}^2 = (\pi / p_{it}^2)[\gamma_{ij} + S_{it}(S_{it} - 1)], \, i = j. \tag{4.11}$$

To satisfy the standard, static economic theory of profit maximization, this matrix of second derivatives should be positive semidefinite at all observations in the data set, $t = 1, T$. The simplest way to check this condition is by examining the eigenvalues of the matrix; if they are all non-negative, the condition is satisfied. Denoting the eigenvalues of the profit function's Hessian by λ_j, $j = 1, J$, a prior for the random parameters of the profit function system of equations in (4.8) and (4.9) can be written as

$$p(\eta, \Sigma) \propto p(\alpha, \beta, \gamma, \theta, \delta, \Sigma) = |\Sigma|^{-(J+1)/2} \left[\prod_{i=1}^{q} N(\eta_i, 1000) \right] D(\lambda_{\min}),$$

$$(4.12)$$

where η is a $(q \times 1)$ vector holding all the independent random parameters (those not identically determined from symmetry or homogeneity relationships to other parameters) from equations (4.8) and (4.9), Σ is the $(J \times J)$ variance-covariance matrix for the error terms from the profit function and the $(J - 1)$ netput share equations that are included in the estimation (remembering that one is dropped), and $D(\lambda_{\min})$ is an indicator function equal to 1 when the minimum eigenvalue for all T observations is non-negative and equal to 0 otherwise. In (4.12), the standard Jeffreys prior has been included for the variance-covariance matrix Σ.

The general steps to performing a Bayesian analysis of this problem using importance sampling follow the pattern from the preceding demand equation example. However, for this application, imagine that the error variance parameters are of interest and the researcher does not wish to integrate the variance-covariance matrix out. If the errors are assumed multivariate normal, the joint posterior distribution for (η, Σ) will take the form

$$p(\eta, \Sigma) \propto |\Sigma|^{-(T+J+1)/2} N(\eta, 1000 I_q) \, D(\lambda_{\min})$$

$$\times \exp \left[-\frac{1}{2} \sum_{t=1}^{T} (z_t - X_t \eta_t)' \Sigma^{-1} (z_t - X_t \eta_t) \right], \quad (4.13)$$

where z_t is a vector of π_t and the S_j, η is a vector of all the free model parameters to be estimated, and X_t is the conformable matrix so that z_t, η, and X_t translate the equations in (4.8) and (4.9) into matrix form.

The presence of a very weakly informative truncated prior for the η parameters causes this posterior distribution to be nonstandard and difficult to generate direct Monte Carlo draws from. To generate samples from a substitute density, choose the posterior distribution divided by the prior on η, as this has the standard form of a multivariate normal-inverted Wishart joint distribution. Geweke (1988) presented the steps to draw random variables from such a distribution in a two-stage process. First, generate a random draw on Σ from its marginal density (an inverted Wishart). Then generate a draw on η from a multivariate normal distribution conditioned on the just-generated random value of Σ. Treat these draws as one observation in the posterior sample space of the parameters, $(\eta^{(i)}, \Sigma^{(i)})$, repeat this process for some large number of times B, and follow the other steps in an importance sampling algorithm, as laid out in the preceeding demand equation application.

The steps to generate the draws for $\Sigma^{(i)}$ following Geweke (1988) are listed here:

1. Calculate the sum-of-squared errors matrix S from the errors of the J estimated equations.
2. Compute the lower triangular Cholesky decomposition of S^{-1}, L such that $LL' = S^{-1}$.
3. Build the random ($J \times J$) lower triangular matrix $U^{(i)}$ where the off-diagonal elements are standard normal random variables, $N(0, 1)$, and the ith diagonal element is a random variable distributed $\chi^2(T - k - i + 1)$ where k is the number of regressors per equation. For an unbalanced model such as this one (with different numbers of regressors in each equation), use the average number of regressors. Random variables can be generated from the $\chi^2(p)$ distribution by generating p independent $N(0, 1)$ random variables, squaring each one, and then summing them.
4. Compute $R^{(i)} = (LU^{(i)})^{-1}$ and then $\Sigma^{(i)} = R^{(i)\prime}R^{(i)}$.

To then generate random draws for $\eta^{(i)}$ conditional on $\Sigma^{(i)}$, follow these steps:

1. Construct the variance-covariance matrix of the η vector as a function of $\Sigma^{(i)}$, denote this matrix by $V^{(i)}$. Compute its lower triangular Cholesky decomposition $C^{(i)}$, $V^{(i)} = C^{(i)}C^{(i)\prime}$.
2. Draw a ($q \times 1$) vector of $N(0, 1)$ random variables, $h^{(i)}$.
3. $\eta^{(i)} = h + C^{(i)}h^{(i)}$, where h is the ($q \times 1$) vector of the mode of the marginal likelihood function for η from which you are drawing (the maximum likelihood estimates).

In the construction of the importance weights, the posterior distribution must be evaluated for each of the B draws. To do this, the researcher will need to build the Hessian of the profit function for that ith draw and for all T observations in the data set and then perform T checks of the convexity condition in order to determine the value of the indicator function $D()$ in the prior distribution. If the value of the prior separate from this indicator function is computed for each draw, the posterior support for convexity in this data set can be calculated using formula (4.5). Posterior means, medians, and interquartile ranges can be computed as described earlier for the parameters in both η and Σ. Such posterior values can also be computed for functions of the parameters that may be of interest, such as elasticities. The hypotheses of homogeneity and symmetry cannot be tested easily in this framework because they were imposed through the dropping of one netput equation and the recovery of those missing parameters.

One of the advantages to imposing curvature conditions through the Bayesian approach of restricting the prior support for the parameters is that the posterior estimates of the parameters will not lie on the cusp between satisfying and violating the restrictions from economic theory. When inequality restrictions are imposed in a maximum likelihood framework, if the mode of the likelihood function lies outside the admissable region, the maximum likelihood estimates will generally be at a point where the restrictions are satisfied with equality (that is binding). Taking a Bayesian approach, the posterior mean and median estimators will lie within the region of positive prior support, not on the boundary of the region. This seems more pleasing from a theoretical point of view.

Other potential applications for imposing sign or curvature conditions that derive from economic theory include estimation of single-equation supply or demand models, utility functions, risk-aversion coefficients, cost functions, demand systems (a common application in the agricultural economics literature), and certain trade models where the Marshall-Lerner condition provides a sign restriction on the sum of specific elasticities. I am certain that readers can lengthen this list considerably.

References

Ball, V. E. (1988). "Modeling supply response in a multiproduct framework." *American Journal of Agricultural Economics* 70, 813–825.

Chalfant, J. A. (1993). "Estimation of demand systems using informative priors." *American Journal of Agricultural Economics* 75, 1200–1205.

Chalfant, J. A., R. S. Gray, and K. J. White (1991). "Evaluating prior beliefs in a demand system: The case of meats demand in Canada." *American Journal of Agricultural Economics* 73, 476–490.

Geweke, J. (1986). "Exact inference in the inequality constrained normal linear regression model." *Journal of Applied Econometrics* 1, 127–141.

Geweke, J. (1988). "Antithetic acceleration of Monte Carlo integration in Bayesian inference." *Journal of Econometrics* 38, 73–89.

Hayes, D. J., T. I. Wahl, and G. W. Williams (1990). "Testing restrictions on a model of Japanese meat demand." *American Journal of Agricultural Economics* 72, 556–566.

Lau, L. J. (1976). "A characterization of the normalized restricted profit function." *Journal of Economic Theory* 12, 131–163.

Lau, L. J. (1978) "Testing and imposing monotonicity, convexity, and quasi-convexity constraints." Appendix A.4 in *Production Economics: A Dual Approach to Theory and Applications*, vol 1., eds. M. Fuss and D. McFadden. Amsterdam: North Holland.

Lopez, R. E. (1984). "Estimating substitution and expansion effects using a profit function framework." *American Journal of Agricultural Economics* 66, 358–367.

Shumway, C. R. (1983). "Supply, demand, and technology in a multiproduct industry: Texas field crops." *American Journal of Agricultural Economics* 5, 748–760.

5

Studying Parameters of Interest

This chapter focuses on inference for functions of the structural parameters. Defining the structural parameters as those that are directly observable in the regression model as specified, the aim of this chapter is to present methods for specifying prior distributions and subsequently deriving posterior distributions for functions of the structural parameters that may be of more interest to the researcher or decision makers. Examples of nonstructural random variables that economists are interested in include elasticities, welfare measures, impulse response functions, the length of the business cycle, and returns to scale.

An Application to Estimating Welfare Measures

Applied studies of international trade issues, publicly funded projects, and government policy changes often rely on estimates of producer and consumer surplus to determine the economic benefits and costs of the issue being researched. These studies typically have several equations (usually supply and demand) that are treated as simultaneous with a set of endogenous variables that are jointly determined (often prices and quantities). Estimates of the structural equation parameters are then used to calculate welfare measures such as producer and consumer surplus. Even though some researchers go through the effort of computing standard errors for the welfare measures (using a bootstrap approach or the delta method), the welfare measures are usually very nonlinear functions of the structural parameters so that the distribution of the welfare measures is likely to be asymmetric. Further, the welfare measures may well be correlated. In such cases, constructing the full posterior distribution of the welfare measures and, perhaps, their difference or ratio can be a great advantage. This is straightforward using a Bayesian approach.

Kloek and van Dijk (1978) first advanced the notion that numerical Bayesian methods were well-suited to attacking economic policy questions within a simultaneous equations framework, proposing importance sampling as a computationally efficient method for estimating the posterior moments of parameters of interest. They also specifically demonstrate the ability to use prior information on nonstructural parameters. In their application to a three-equation system, they derive posterior moments for the structural parameters, reduced form parameters, and a long-run multiplier. This is very similar to the following application, which calculates posterior distributions of welfare measures; researchers interested in this type of application are strongly encouraged to read Kloek and van Dijk's classic paper. Other useful references for Bayesian estimation of systems of simultaneous equations include Drèze and Richard (1983), Richard (1973), Zellner (1971, ch. 9; 1978), Zellner and Park (1979), and Zellner, Bauwens, and van Dijk (1988).

For a linear simultaneous equations model that generally follows the presentation of Kloek and van Dijk (1978), Bayesian analysis of nonlinear functions of the structural parameters can proceed as follows. Begin with the model in matrix notation

$$Y\Gamma + XB = U \tag{5.1}$$

where Y is an $(n \times m)$ matrix of observations on the endogenous variables, X is an $(n \times k)$ matrix of exogenous variables, U being an $(n \times m)$ matrix of error terms, and Γ and B are conformable matrices of parameters to be estimated. Given standard model identification assumptions and time-independent multivariate normality of the error terms, the likelihood function for this model is

$$p(Y|Z, \Gamma, B, \Sigma) \propto |\Sigma|^{-n/2}\|\Gamma\|^n$$
$$\times \exp(-0.5tr\{[\Gamma'\hat{S}\Gamma + (B - \hat{B})'X'X(B - \hat{B})]\Sigma^{-1}\}) \tag{5.2}$$

where Σ is the covariance matrix of the error terms, $\hat{S}$ is the sum-of-squared errors matrix from the reduced form equation $Y = Z\Pi + V$, and $\hat{B} = -\Pi\Gamma$.

To construct a prior distribution for the parameters (Γ, B, Σ), divide them into three subsets: the covariance matrix Σ, which will almost always be modeled with a standard Jeffreys prior $p(\Sigma) \propto |\Sigma|^{-(m+1)/2}$; some subset ζ of the matrices Γ and B that are not random parameters due to (identification-related) restrictions, generally equal to either 0 or 1, these are not really parameters and will be treated as constants to condition upon in deriving the distributions of the other parameters; and, finally, the remaining parameters of Γ and B, denoted by the vector θ, about which we possess (at least weakly) informative prior information. Write this informative prior as $p(\theta)$, so that the full prior distribution for the parameters (Γ, B, Σ) is

$$p(\Gamma, B, \Sigma) = p(\theta, \zeta, \Sigma) \propto p(\theta)|\Sigma|^{-(m+1)/2}. \tag{5.3}$$

The easiest way to analyze the posterior distribution of θ or some function $g(\theta)$ is to work with the marginal posterior distribution of θ. If we integrate out Σ and condition on the known parameters ζ (related to identification), the marginal posterior distribution of θ is

$$p(\theta|Y, X, \zeta) \propto p(\theta)\|\Gamma\|^n|\Gamma'\hat{S}\Gamma + (B - \hat{B})'X'X(B - \hat{B})|^{-n/2}. \tag{5.4}$$

To analyze this distribution numerically, or to build a numerical approximation to the marginal posterior distribution of some $g(\theta)$, use importance sampling. The marginal posterior distribution in equation (5.4) is related to a multivariate Student-t distribution, with the first two terms causing the discrepancy. Thus, a multivariate

Student-t distribution with a small degree-of-freedom parameter centered at standard 3SLS estimates of Γ and B is a good candidate for the substitute density. Random draws on Γ and B can be generated from the substitute density, welfare measures and the value of the marginal posterior can be computed for each draw, saved, and posterior means, medians, and interquartile ranges can be computed as outlined in previous examples. This allows for Bayesian analysis of interesting policy questions within a simultaneous equations framework even when one has prior information on some parameters or functions of parameters.

An Application to Elasticities

In some models, elasticities are nonlinear functions of the structural parameters. In such cases, not only are these nonstructural parameters important as policy-relevant variables but they are also a parameter space in which economists are more likely to have informative prior information. When using numerical methods, it is possible to use a prior distribution that is specified in terms of a different parameter space from the the likelihood function. The likelihood function is almost always measured with respect to the structural parameters, but sometimes the researcher has prior information that is more easily expressed in terms of a function of the structural parameters. Elasticities are a good example of such a situation.

Take the translog profit function of Chapter 4 as the application again here, but with a different prior distribution. The own-price elasticities of the input demand and output supply (netput) equations take the form

$$\xi_{ii}(t) = (\pi/p_{it}S_{it})[\gamma_{ii} + S_{it}(S_{it} - 1)] \tag{5.5}$$

and the cross-price elasticities have the formula

$$\xi_{ij}(t) = (\pi/p_{it}S_{it})[\gamma_{ij} + S_{it}S_{jt}]. \tag{5.6}$$

If the observed values of the profit shares and profit are used in equations (5.5) and (5.6), then the elasticities are simply linear functions of the random parameters γ_{ij} and their distributions can be easily computed from the distribution of the γ_{ij}. However, it is common practice to use estimated (i.e., fitted) share and profit values in the elasticity formulas. This turns the elasticities into highly nonlinear functions of the model's structural parameters and results in a distribution for the elasticities that more accurately reflects the true uncertainty about these parameters. The Bayesian approach, with reliance on numerical methods, allows for the posterior distributions of the elasticities to be derived without any great difficulty. In fact, it is easy to specify the prior distribution for the parameters in terms of the elasticities ξ instead of the structural parameters ($\alpha, \beta, \gamma, \theta, \delta$). Because all the structural parameters appear in the model's predicted values of the profit function and most appear in some of the share equations, a prior over the elasticities ξ implies a prior distribution over the structural parameters (although the mapping may be quite complex and may not be unique). This should not cause a problem with the calculation of posterior distributions as long as the researcher ensures (or checks, *ex post*) that the prior distributions are well-behaved (proper, with at least two moments) for the parameters whose posterior distributions they build and the Jacobian for the change of variables is included in the formulas for computation of posterior means, medians, and so on. In the nomenclature of this paragraph, the Jacobian is simply the determinant of the matrix of partial derivatives of the elasticities ξ with respect to the structural parameters $(\alpha, \beta, \gamma, \theta, \delta)$.

Economists generally have little intuition of the magnitude, or even sign, of many of the structural parameters in a translog profit function; however, we often have well-formed opinions about supply and demand elasticities, both own- and cross-price. We expect the magnitudes to be rather small; for own-price elasticities we expect a particular sign, and at times we even have prior beliefs about whether the elasticity is greater or less than 1 in absolute value (elastic or inelastic). A simple prior on the vector of elasticities ξ would be the product of independent univariate normal distributions centered at either 1 or -1 depending on the expected sign of each elasticity with variances equal to 5 or 10. This provides a good bit of prior support for elasticity values away from unitary elasticities and does not impose with certainty any sign restrictions. A rather diffuse prior such as this is a good, cautious one for a situation in which the prior distribution on the structural parameters is not directly observable.

Examining the posterior distributions for the elasticities would proceed by importance sampling with the following steps:

1. Specify the prior distribution for the elasticities (in place of one on the structural parameters), along with the standard Jeffreys prior on the covariance matrix of the error terms.
2. Generate draws on Σ and the structural parameters from a substitute density using the steps outlined in Chapter 4's discussion of the translog profit function (Geweke, 1988a).
3. Compute the values of the substitute density, the elasticities ξ, the prior distribution, the Jacobian, the likelihood function (from the form contained in equation (4.13)), and any other functions of interest for each draw, $i = 1, B$. B is the number of draws and should be at least 10,000 and hopefully 20,000 for an application like this, which studies highly nonlinear functions of the structural parameters using importance sampling and a prior on nonstructural parameters.
4. Using the appropriate formulas for importance sampling applications, compute posterior means, medians, and other percentiles desired for the elasticities and other functions. The standard formulas presented previously are modified by the addition of the Jacobian, which multiplies the prior on the elasticities everywhere the prior appears in a formula. Make sure to check the numerical approximation errors for the elasticities.
5. As an additional check on the procedure's success, build the empirical prior distributions of some or all of the structural parameters. If the drawn values of the structural parameters are saved, the prior distribution's value for each draw can be substituted in place of the posterior distribution in the standard importance sampling formula for computing means, medians, and percentiles

of a distribution. In this manner, you can recover a "picture" of the implied prior distributions of the structural parameters. This provides a check that these implied prior distributions are well-behaved and that the posterior distributions will exist and have sufficient moments for empirical investigation.

It is also possible to analyze elasticities under prior distributions that truncate prior support to a parameter space that imposes signs for the elasticities suggested by economic theory. Refer to the techniques discussed in Chapter 4 for methods of working with this class of prior distributions.

An Application to Dynamic Properties

Other opportunities to apply Bayesian methods to the study of nonstructural parameters abound in applied economics. Geweke (1988b) investigated the length of the business cycle in OECD countries. He used a Bayesian approach to derive posterior estimates of the cycle length implied by the coefficients of an autoregressive model of GNP or GDP for each country. While Geweke did not use an informative prior, he could have, and this sort of application is a natural for numerical Bayesian techniques. As the final example of this chapter, the steps to analyzing the cycle length of a time series variable with an informative prior are laid out. This is a slight modification of Geweke (1998b) that incorporates an informative prior distribution on the nonstructural parameters of interest.

Begin with a univariate autoregressive model for some economic time series variable y_t,

$$y_t = \mu + \rho_1 y_{t-1} + \rho_2 y_{t-2} + \rho_3 y_{t-3} + \varepsilon_t, \tag{5.7}$$

where ε_t is an iid normally distributed innovation (error) term and μ and the ρ_i are unknown parameters. We are not directly interested in the structural parameters, but in the cycle length of this autoregressive process (and on the probability that a cycle is indeed part of the series' dynamic behavior). The length of cycle implied by the ρ_i parameters (μ is not involved) is given by

$$\omega = 2\pi / tan^{-1} [Im(\lambda_c)/Re(\lambda_c)], \tag{5.8}$$

where λ_c is one eigenvalue from a complex conjugate pair of the three eigenvalues from the matrix

$$A = \begin{bmatrix} \rho_1 & 1 & 0 \\ \rho_2 & 0 & 1 \\ \rho_3 & 0 & 0 \end{bmatrix}. \tag{5.9}$$

To analyze the posterior distribution of the cycle length ω, again rely on importance sampling to allow for an informative prior on ω. Assume the standard Jeffreys prior for the variance of ε_t, σ^2, and an uninformative prior for the location parameter μ. Because the eigenvalues are functions of all three ρ_i parameters, we can express the prior distribution in terms of (μ, σ, ω). If the data is quarterly U.S. real GNP, one might choose a prior such as

$$p(\mu, \sigma, \omega) \propto N[(\omega - 20)/6]/\sigma, \qquad (5.10)$$

where $N[]$ represents the standard normal distribution. This places the mode of the prior distribution for cycle length at five years while providing support for a reasonable range of cycle lengths, it only places a very negligible portion of the prior support on the nonsensical nonpositive region of the parameter space for ω. To create a square Jacobian, create a vector of three variables, $\zeta = (\lambda_1, \lambda_1^2, \omega)$ where λ_1 is the third eigenvalue (the one that is not part of the complex conjugate pair). Then, the Jacobian is $J = |\partial\zeta/\partial\rho|$ where ρ is the vector of the three ρ_i. In this application, numerical evaluation of the nine derivatives involved in the Jacobian is the best way to compute its value.

The likelihood function for this model is

$$p(y|\mu, \sigma^2, \rho, y_0, y_{-1}, y_{-2}) \propto \sigma^{-T}$$
$$\times \exp[-0.5(y_t - \mu - \rho_1 y_{t-1} - \rho_2 y_{t-2} - \rho_3 y_{t-3})^2/\sigma^2)], \qquad (5.11)$$

where the likelihood function is conditioned on some initial conditions for the time series variable and T is the size of the sample (net of these initial conditions). With this prior and likelihood function, the marginal posterior distribution of $(T - 4)s^2/\sigma^2$ is $\chi^2(T - 4)$, the conditional posterior distribution of μ is normal and the conditional distribution of the ρs is a product of two normal distributions (the prior and the kernel of the likelihood function). So taking an importance sampling approach, we can follow Geweke's approach to generate draws from a substitute density.

1. First draw a value of $1/\sigma^{2(i)}$ by generating a random draw from the $\chi^2(T - 4)$ distribution and scaling it by s^2; then invert the value to produce a random draw for $\sigma^{2(i)}$. Then draw μ and the ρs from the multivariate normal distribution of the maximum likelihood estimates except for replacing the maximum likelihood estimate of σ^2 with $\sigma^{2(i)}$.
2. Compute $\omega^{(i)} = g[A(\rho^{(i)})]$ using equations (5.8) and (5.9). If there are no complex eigenvalues, set the value of $\omega^{(i)} = 0$.
3. Compute the value of the prior distribution for $\omega^{(i)}$, $N[(\omega^{(i)} - 20)/6]$. For this substitute density, the prior on the cycle length is the importance weight. Compute the Jacobian, $J^{(i)}$, by evaluating the partial derivatives of the three variables in ζ with respect to ρ at the drawn value of $\rho^{(i)}$.
4. Using the simplified importance weight for this application, the numerical approximation to the posterior mean of the cycle length is computed as

$$\omega^{(i)} = \frac{\sum_{i=1}^{B} \omega^{(i)} N[(\omega^{(i)} - 20)/6] J^{(i)}}{\sum_{i=1}^{B} N[(\omega^{(i)} - 20)/6] J^{(i)}} \qquad (5.12)$$

5. Posterior medians, percentiles of the posterior distribution of ω, and the standard deviation of the numerical approximation error can also be estimated by the formulas for importance sampling algorithms with the importance weights specified earlier.

Table 1.   Posterior Estimates of the Length of the Business Cycle

Mean	Median
18.53	18.87
(0.08)	(16.75, 21.14)
	(14.25, 23.80)

The results are based on 10,000 draws. The number in parenthesis under the posterior mean is the standard deviation of the numerical approximation error. Those under the posterior median are the interquartile range and the limits of a 90% highest posterior density region.

6. By recomputing the posterior estimates under several alternative priors with different prior means and variances, the sensitivity of the results to the prior distribution can be established.
7. By constructing the ratio of the sum, the importance weights for all draws with nonzero cycle lengths to the sum of the importance weights for all B draws, you produce the posterior probability for the existence of a business cycle.

A table of the empirical results from this application might look something like Table 1. Table 2 shows how one might display the results of sensitivity analysis on the prior distribution.

References

Drèze, J., and J.-F. Richard (1983). "Bayesian analysis of simultaneous equation systems," *Handbook of Econometrics, vol* 1, Z. Griliches and M. D. Intriligator, eds. New York: North Holland.

Geweke, J. (1988a). "Antithetic acceleration of Monte Carlo integration in Bayesian inference." *Journal of Econometrics* 38, 73–89.

Geweke, J. (1988b). "The secular and cyclical behavior of real GDP in 19 OECD countries, 1957–1983." *Journal of Business & Economic Statistics* 6, 479–486.

Kloek, T., and H. K. van Dijk (1978). "Bayesian estimates of simultaneous equation system parameters: An application of integration by Monte Carlo." *Econometrica* 46, 1–19.

Richard, J.-F. (1973). *Posterior and Predictive Densities for Simultaneous Equation Models*. Berlin: Springer-Verlag.

Zellner, A. (1971). *An Introduction to Bayesian Inference in Econometrics*. New York: John Wiley & Sons.

Zellner, A. (1978). "Estimation of functions of populations means and regression coefficients: A minimum expected loss approach." *Journal of Econometrics* 8, 127–158.

TABLE 2.   Effect of the Prior Distribution on the Posterior Mean of ω

Prior Mean	Prior Standard Deviation	Posterior Mean
20	6	18.53
18	6	18.11
16	6	17.85
20	8	18.42
20	4	18.72
20	2	18.82

The first row displays the base results. All results are based on the same 10,000 draws.

Zellner, A., L. Bauwens, and H. K. van Dijk (1988). "Bayesian specification analysis and estimation of simultaneous equation models using Monte Carlo methods." *Journal of Econometrics* 38, 39–72.

Zellner, A., and S. B. Park (1979). "Minimum expected loss (MELO) estimators for functions of parameters and structural coefficients of econometric models." *Journal of the American Statistical Association*, 74, 185–193.

6

Unit Root and Cointegration Tests

This chapter can almost be treated as a special subcase of last chapter's topic, the study of nonstructural parameters. The econometrics of unit root and cointegration models, or more generally, the study of the (non)stationarity of time series data, both univariate and multivariate, is at its simplest level a question about the value of a single scalar parameter. For a univariate time series whose stationarity is at issue, the question revolves around the value of the dominant root of the dynamic process: is it less than, equal to, or greater than 1? These equate to the cases of stationarity (order of integration 0), a unit root (integrated order 1), and an explosive nonstationary root (also integrated order 1). In multivariate studies, most questions about the dynamic properties of a set of series can be reduced to inferences regarding a parameter such as the value of the dominant root of a linear combination of the series (a cointegration test) or the difference in the number of roots of magnitude 1 or greater between the multivariate process and its associated set of separated univariate processes (another way to view a cointegration test). In this chapter, I shall briefly review some of the basics of unit root and cointegration testing and present some applications to conducting inference on the corresponding parameters using Bayesian methodologies.

The Basics of Unit Root Tests

A univariate time series is said to have a unit root when one of the roots of the determinental polynomial of the series has magnitude equal to 1 (lies on the unit circle). For a stationary series, all roots should lie outside the unit circle (implying parameters with magnitudes less than 1); explosive roots lie inside the unit circle. A stationary series has a defined expected value, or mean, and is mean-reverting, implying that it tends to return to its central value. Nonstationary series (those with unit or explosive roots) do not have an unconditional expected value, only conditional expected values for a specific time period, conditioned on some initial condition. The statistical distributions of many sampling theory estimators are dependent on the stationarity properties of the time series being modeled, so the testing of series for unit (and explosive) roots has become important over the last

15 years as the distributional theory of stationary and nonstationary time series has become better understood and more fully developed.

The standard sampling theory test for a unit root is the Dickey-Fuller test, which tests the null hypothesis of a unit root versus the alternative hypothesis of a stationary root using a model of the form

$$\Delta y_t = \mu + \delta t + \sum_{s=1}^{p} \Delta y_{t-s} + (\rho - 1)y_{t-1} + \varepsilon_t, \tag{6.1}$$

where y_t is a univariate time series whose stationarity is in doubt, Δ is the difference operator, $\Delta y_t = y_t - y_{t-1}$, t denotes time periods, μ and δ are parameters that allow for trend and drift, ρ is the root whose value is at issue, and ε_t is a white noise iid error term. A researcher using a model such as that in (6.1) might perform an augmented Dickey-Fuller (ADF) test where the null hypothesis is $\rho = 1$, the alternative hypothesis is $\rho < 1$, and the word augmented refers to the inclusion of additional lags of the time series variable y to allow for more general and complex dynamic properties than a simple random walk. The test is performed by constructing the standard t-value for such a test, but using critical values from special Dickey-Fuller test tables that account for the nonstandard distribution on the test statistic under the null hypothesis of a unit root (Dickey and Fuller, 1979). Other tests have been developed that allow for heteroscedasticity (Phillips and Perron, 1988), unknown lag length (Dickey and Pantula, 1987), other parameterizations of the trend and drift terms that are more stable under both null and alternative hypotheses (Schmidt and Phillips, 1992), and other variations on the basic Dickey-Fuller approach.

A big problem with this approach is that the null hypothesis is that of a unit root and the sampling theory approach forces the alternative hypothesis to meet a large burden in order to force a rejection of the null hypothesis. The two competing hypotheses are not treated equally and the posterior probability (likelihood value) of the null is not even considered in the testing procedure. This leads to very low power for such tests, the inability to reject the null hypothesis of unit roots when no unit root exists (DeJong et al., 1992). A Bayesian test for stationarity vs. nonstationarity focuses directly on the relative posterior probabilities (the posterior odds ratio) of the two hypotheses.

To facilitate this focus on the posterior distribution of the dominant root (the one suspected of being nonstationary), it is useful to use the relation referred to in Chapter 5 during the discussion of Geweke's study of the length of the business cycle (Geweke, 1988b). Start with a univariate autoregressive model for some time series variable y_t,

$$y_t = \mu + \rho_1 y_{t-1} + \rho_2 y_{t-2} + \rho_3 y_{t-3} + \varepsilon_t, \tag{6.2}$$

where ε_t is an iid normally distributed innovation (error) term and μ and the ρ_i are unknown parameters. The dynamic properties of this model can be investigated easily by direct examination of the matrix

$$A = \begin{bmatrix} \rho_1 & 1 & 0 \\ \rho_2 & 0 & 1 \\ \rho_3 & 0 & 0 \end{bmatrix}. \tag{6.3}$$

The largest eigenvalue of the matrix A in equation (6.3) is the dominant root for the model in equation (6.2) and the root we suspect of nonstationarity. Thus, it is the maximum eigenvalue of A around which we should build a Bayesian unit root test. (Note that working with a more general AMRA(p, q) model would not complicate the investigation at all except for adding to the dimension of the numerical integration involved; the form of the A matrix is unaffected by the presence of moving average terms). Further, in keeping with the example set in the last chapter, I recommend that the prior distribution for the model in equation (6.2) be expressed in terms of the eigenvalues of A (along with a prior on the variance of the error terms). One should be very cautious in so doing, as a recent controversy has erupted over the fact that a flat, seemingly diffuse prior in the parameter space of A's eigenvalues does not imply a flat prior on the structural parameters of equation (6.2), the ρ_i, and vice versa (cf. Phillips, 1991, with extensive discussions). Remember to include the Jacobian in any computations where the draws are generated for the ρs and the prior is specificied in terms of the eigenvalues of A.

This linkage is a mathematical fact due to the functional relationship between the ρ_is and the eigenvalues of A, but it should not be treated as a hindrance. I believe this fact makes even more clear the advantage of specifying the prior distribution in terms of the eigenvalues about which we tend to be reasonably well-informed relative to the structural parameters ρ_i. Phillips (1991) showed that a flat (Jeffreys) prior on the ρ_i implied an improper prior on the dominant root that is increasing at an increasing rate. He also derived the Jeffreys prior on the dominant root and found this to have an exponential shape that is also sharply increasing in the range of parameter values of most interest (those close to unity). Neither prior seems well-suited to the task of performing unit root tests without a prior that will be offensive to interested readers of the empirical results. Thus, I suggest that researchers stick to proper, informative priors for their Bayesian unit root tests. This should not be too controversial, because if there was not a fair amount of agreement that the dominant root was somewhere in the statistical neighborhood of unity we would not be thinking of performing a test to check for stationarity.

In previous work on this topic, I have used a beta prior on the eigenvalues of the A matrix due to the convenient properties of the beta distribution and the ease of working analytically with the prior at the beginning of the research process when the prior is specified (Dorfman, 1993). The advantages of the beta distribution are its natural unit range from 0 to 1 (which can be easily shifted or scaled to move or expand the range of values with positive prior support), its flexibility from diffuse to highly informative, and its reliance on only two parameters (α, β). A beta distribution, Beta(α, β), has mean $\alpha/(\alpha + \beta)$ and mode $(\alpha - 1)/(\alpha + \beta - 2)$. The Beta$(1, 1)$ is a uniform distribution with range $[0, 1]$; as the two parameters increase, the density's shape becomes more peaked (informative) and can approach the shape of the normal and Student-t distributions. For $\alpha > \beta$, the distribution is skewed to the right; for $\alpha < \beta$, the distribution is skewed to the left. In the first application here, a multivariate beta prior on the roots of an AR(3) model will be presented and used to conduct unit root tests.

Once a prior on the roots (and variance of the error terms), the Jacobian for the change of variables, and the likelihood function have been specified, the researcher is ready to conduct a test for the stationarity of a time series. A Bayesian test for stationarity versus nonstationarity of a time series can be conducted as a posterior odds ratio test. For concreteness, assume the AR(3) model of equation (6.2), denote the three eigenvalues of the A matrix defined by (6.3) using the symbols λ_i, $i = 1$, 2, 3. Finally, represent the magnitudes of these three eigenvalues by Φ_i, $i = 1, 2$, 3, with the Φ_i ordered largest to smallest. The odds ratio of interest will be

$$K_{12} = \frac{\int_0^{1-\varepsilon} p(\Phi_1|y)\,d\Phi_1}{\int_1^{\Phi_{max}} p(\Phi_1|y)\,d\Phi_1} \tag{6.4}$$

where $p(\Phi_1|y)$ is the marginal posterior distribution of the dominant root, ε is an arbitrarily small positive number, and Φ_{max} is the largest value of Φ_1 that receives positive prior support. This marginal posterior distribution of the dominant root is given by

$$p(\Phi_1|y) = \iiint p(\Phi_1, \Phi_2, \Phi_3, \sigma)\, p(y|\Phi_1, \Phi_2, \Phi_3, \sigma)\,d\Phi_2\,d\Phi_3\,d\sigma \tag{6.5}$$

where the joint posterior is shown as the product of the prior distribution and the likelihood function and the effects of the other three parameters are integrated out. If the posterior odds ratio in equation (6.4) is greater than 1, the posterior distribution favors stationarity; if the ratio is less than 1, the posterior distribution favors nonstationarity (a unit or slightly explosive root). The exact value of the posterior odds ratio necessary to support or reject one of the two hypotheses concerning the dynamic properties of this time series is selected by the researcher through the specification of their loss function over these decisions. Whether to deviate from a balanced loss function that makes decisions based on a dividing line of unit posterior odds, and in which direction, depends on the existing theories covering the application at hand. When theory strongly suggests that a series should be nonstationary, one might continue to support that proposition until the posterior odds ratio exceeded 2 or 3 (representing posterior support of stationarity of 0.67 or 0.75, respectively).

Note that the posterior odds ratio test is done for stationarity versus nonstationarity, with the nonstationary region covering the parameter space from an exact unit root to a slightly explosive root. The size of the region entertained for the nonstationary hypothesis is up to the researcher. Including roots that are at least a little above unity in a hypothesis of nonstationarity offers advantages over simply testing stationarity relative to an exact unit root as certain technical problems occur when using a posterior odds ratio test for a point hypothesis against a competing diffuse hypothesis. In particular, if a point hypothesis of a unit root is specified, the prior distribution must be adjusted so that the point hypothesis does not receive zero prior support (because a single point in a continuous parameter space has probability 0). Thus, I recommend that whenever possible Bayesian tests be conducted as stationarity versus nonstationarity with each hypothesis encompassing a well-defined parameter space that contains more than a single point in the range of the dominant root. If a researcher wants to test an exact unit root versus the stationary alternative, one must work with the marginal posterior of the unit root parameter so that an integration step is still necessary to remove the effect of conditioning on any extraneous parameters. For a great discussion of the theory involved in unit root testing see Schotman and van Dijk (1991b), which is part of the *Journal of Applied Econometrics* special issue on unit root tests highlighted by Peter Phillips' (1991) paper on Bayesian unit root tests with its emphasis on prior distributions.

An Application to Efficient Market Tests

Many studies have been done, mainly using sampling theory approaches, to investigate various asset price time series for unit roots. Under a rather simplistic view of such asset markets, if the price series has a unit root, then the asset market is deemed efficient because the future path of the series cannot be accurately forecasted. While a set of simplifying assumptions is necessary to reduce a test of efficient markets to a test for a unit root (or stationarity versus nonstationarity), such applications are performed and provide an empirical base of published studies on performing unit root tests. Using some results from a published study on futures contract price data for concreteness, one such application is presented here.

Dorfman (1993) examined hourly corn and soybean futures prices from 1990 in subsets of 300 observations; here we will focus on his results from the corn data. The test for nonstationarity has four steps: specification of the two hypotheses, specification of the prior, specification of the likelihood, and computation of the posterior odds ratio. The two competing hypotheses were specified as H_1: stationarity ($\Phi_1 < 1.00$) and H_2: nonstationarity ($1.0 \leq \Phi_1 \leq 1.03$). Thus, a slightly explosive dominant root is allowed for (although in retrospect the upper limit is probably too high for data with such a high frequency of observation). This explosive region of the posterior distribution can also be thought of as posterior support for a unit root that is slightly shifted due to sampling error. The model chosen to approximate the data-generating process for both hypotheses was an AR(3) with two additional exogenous variables: an intercept and a linear time trend.

In many economic applications, the researcher has a well-informed prior on Φ_1, with most prior support concentrated in an area "near" unity, but much less information about the magnitude of the smaller roots. It may also be useful to allow the dominant root to take slightly explosive values (greater than 1), although not much greater than 1 as models that are rapidly explosive become obviously nonstationary by ordinary observation and do not necessitate an econometrician to determine their time series properties. A multivariate beta prior for the roots that meet these conditions can easily be constructed as a product of univariate beta distributions (a simple version of the multivariate beta). For the application in Dorfman (1993), prior distributions for the two smaller roots were specified as Beta(1.1,1.1) distributions that look like flat, rounded hills. The prior on the dominant root was specified as a Beta(30,2) for the mean-shifted variable (Φ_1 − 0.03), giving positive prior support over the range $\Phi_1 \in [0.03, 1.03]$. This prior distribution is sharply skewed to the right, has a prior mean of 0.9675, and a prior mode of 0.9667. A standard Jeffreys prior is taken for the variance, allowing for easy analytical derivation of the marginal posterior distribution of the roots. For examining the sensitivity of the empirical results to the prior specification, all odds ratios were also computed under a flat prior on the three dominant roots.

In Dorfman (1993), the tests were performed under two different likelihood function specifications to examine the impact of the likelihood function on the results of the test. First, a nonparametric density was chosen, using a Gaussian kernel function that can be written for a single observation's error term as

$$p(e_t) = \frac{c}{Th} \sum_{i=1}^{T} \exp\left[\frac{-(e_t - e_i)^2}{2h^2}\right] \tag{6.6}$$

where T is the number of observations (300), $c = (2\pi)^{-\frac{1}{2}}$, and $h = 1.66444\sigma T^{-1/5}$ is the bandwidth. The alternative likelihood function specification was Gaussian, assuming that the errors of the AR(3) model are iid normal random variables with zero mean. Thus, four sets of posterior odds ratios were computed, pairing each of the two likelihoods with each of the two prior specifications (nonparametric and Gaussian likelihoods, beta and flat priors).

The posterior odds ratios were computed using importance sampling on the marginal posterior distribution of the roots. Draws on the three autoregressive parameters are made from a trivariate normal distribution centered at the least squares estimates and with metric equal to the least squares covariance matrix scaled by 1.5. For each draw, the three roots can be found by solving for the eigenvalues of the A matrix displayed in equation (6.3). With the value of $\Phi^{(i)}$, the Jacobian, and each of the prior distributions, and the likelihood functions can be evaluated easily and the posterior support for each of the two hypotheses can be computed using the formula for importance sampling. If an indicator function $D(\Phi^{(i)})$ is defined that equals 1 when the dominant root satisfies H_1 and equals 0 when the dominant root satisfies H_2, the posterior probability in support of H_1 is given by

$$p(H_1|y) = \frac{\sum_{i=1}^{B} D(\Phi^{(i)}) p(\Phi^{(i)}) J^{(i)} p(y|\Phi^{(i)})/g(y|\Phi^{(i)})}{\sum_{i=1}^{B} p(\Phi^{(i)}) J^{(i)} p(y|\Phi^{(i)})/g(y|\Phi^{(i)})} \tag{6.7}$$

TABLE 3.   Posterior Odds Ratio in Favor of Stationarity for 1990 Corn Futures Prices

Sample	K_{gb}	K_{gf}	K_{nb}	K_{nf}
1	23.36	31.65	15.80	16.56
2	20.12	29.59	34.97	48.78
3	0.7352	0.6682	0.7265	0.6823
4	17.51	24.88	6.297	6.752

The subscripts denote the likelihood and prior specifications: g for Gaussian, n for nonparametric, b for the beta prior, and f for a flat prior. The results are based on the analysis performed in Dorfman (1993); see the full paper for more details. All posterior probabilities are based on 5000 draws using importance sampling.

where $\Phi^{(i)}$ represents the vector of three roots from the ith draw, B is the number of Monte Carlo draws (5000 in Dorfman, 1993), $p(y|\Phi^{(i)})$ is the likelihood function of the data (either nonparametric or Gaussian), and $g(y|\Phi^{(i)})$ is the substitute density. The posterior support for H_2 can be found by substituting $[1 - D(\Phi^{(i)}]$ in place of $J(\Phi^{(i)})$ in equation (6.7), or by simply taking $1 - p(H_1|y)$. The posterior odds ratio is then computed as $K_{12} = p(H_1|y)/p(H_2|y)$.

The empirical results from Dorfman (1993) are presented as Table 3. They show that for three of the four subperiods, the corn futures market does not appear to be

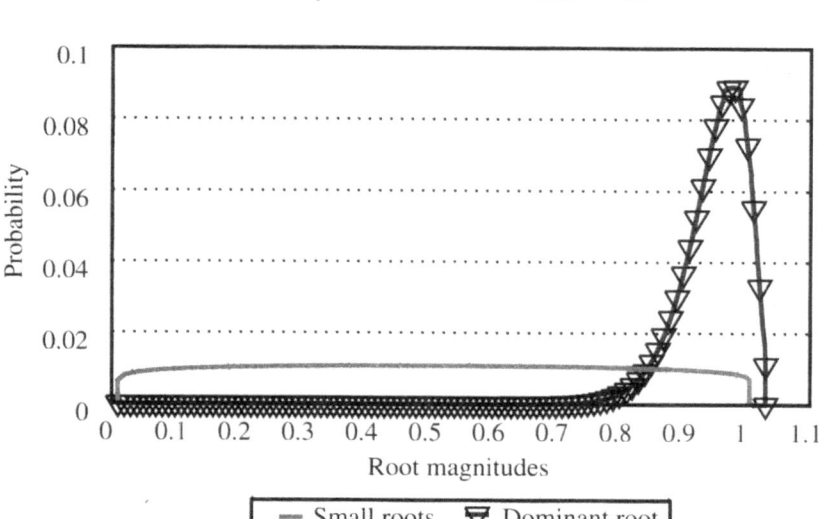

Beta prior distributions on the roots.

FIGURE 4.

Sample 3 posterior distributions of dominant root.

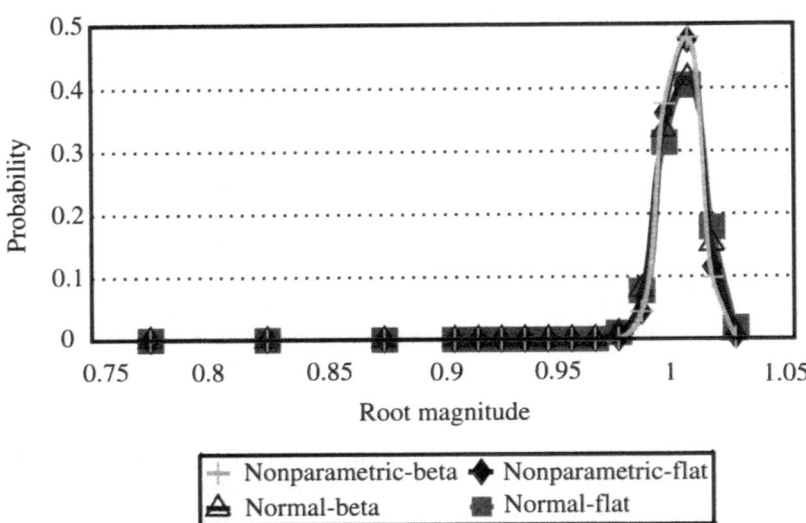

FIGURE 5.

efficient if that is defined as a nonstationary price time series. The posterior odds ratios strongly favor the price series being stationary for all subperiods except the third. Under the nonparametric likelihood specification, the fourth period results are the closest of the other odds ratios to favoring nonstationarity (and the market efficiency that goes with it). These odds ratios imply that the posterior probability of nonstationarity is only 0.12; one would need a loss function strongly penalizing incorrect rejection of efficiency to maintain support for H_2 in the face of these results. To make the evidence clear, graphs of the prior and posterior distributions help. Figure 4 shows the beta prior distribution on the roots. Figure 5 displays the four marginal posterior distributions of the dominant root (under different priors and likelihood specifications) for the third sample, the one that favors nonstationarity according to the posterior odds ratio. Figure 6 shows the same four marginal posterior distributions for the fourth sample, one that has posterior odds ratios strongly favoring stationarity. I believe that looking at these graphs allows the viewer to quickly decide (without computing odds ratios) that the third sample appears to be nonstationary, while the fourth sample appears to be stationary.

Other Applications of Unit Root Tests

The literature on Bayesian unit root tests started with Sims (1988), who developed a very simple test for a random walk versus a stationary AR(1) alternative using a flat prior on the autoregressive parameter in a space from (ρ_{min}, 1) and a discrete mass prior placed on the random walk value of $\rho = 1$. The Bayesian approach to unit root tests was then widely popularized in macroeconometrics by DeJong and Whiteman (1991), who applied slightly more sophisticated priors and investigated the properties of the famous Nelson-Plosser macroeconomic time series. A fabulous collection of articles on Bayesian unit root testing is the special edition of the *Journal of Applied Econometrics* on this topic, which is highlighted by the lead paper by Phillips (1991) and followed by extensive comments, discussion, and his rejoinder. Anyone considering applying Bayesian methodology to tests of time series properties should read this entire volume. Other good applications of Bayesian unit root tests can be found in Koop (1992) and Schotman and van Dijk (1991a).

These papers present a wide variety of prior distributions that can be applied to possibly nonstationary time series models and provide lots of discussion on the pros and cons of different families of prior distributions. They also present both analytical and numerical approaches to computing the posterior distributions and test statistics. However, virtually all these papers rely on a posterior odds ratio test to decide the question of interest. In this sense, the application presented earlier is very representative of the body of literature on this topic.

It is also possible to conduct a posterior odds ratio test for stochastic versus deterministic trends; that is, for the hypothesis of a nonstationary root versus the hypothesis of a stationary root plus a trend (i.e., trend stationarity). Such a test was performed by DeJong and Whiteman (1991) using the posterior odds ratio and defining the region for a nonstationary root to be the interval [0.975, 1.05]. While

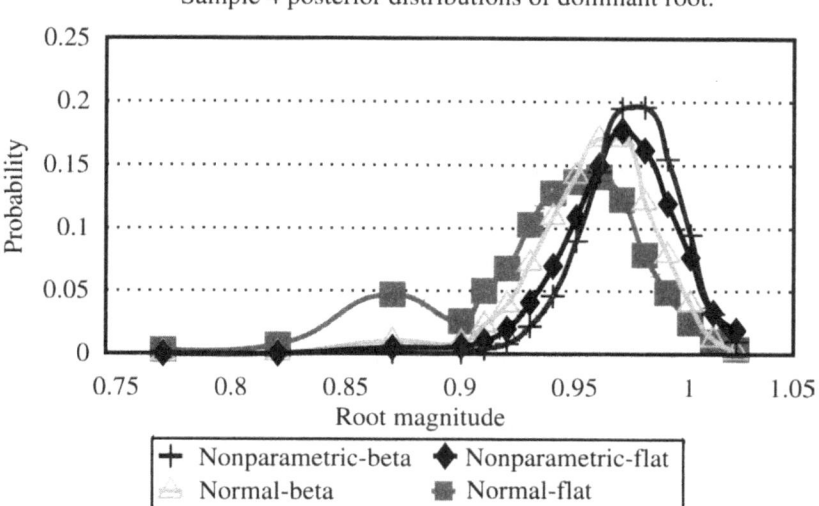

Sample 4 posterior distributions of dominant root.

FIGURE 6.

0.975 is a smaller minimum value for the dominant root than was included earlier in the hypothesis of a nonstationary root, DeJong and Whiteman were working with the Nelson-Plosser macroeconomic data set of annual observations. With that observation frequency, they point out that a root of 0.975 implies shocks have a half-life of 27 years, which they deem pretty close to permanent. Such tests of trend stationarity versus difference stationarity can be accomplished by simple modifications of the procedure outlined earlier for a standard test of a stationary versus nonstationary dominant root.

Bayesian Approaches to Cointegration Tests

Cointegration tests have been conducted in a Bayesian framework by Koop (1991), DeJong (1992), and Dorfman (1995). Their approaches were quite different. Koop (1991) specifies a particular structural model that fits his application to a bivariate time series of stock prices and dividends and constructs a Bayesian test of cointegration based on the parameter restrictions that are implied by the theory of cointegration. This approach is very elegant and should work well for bivariate models when the suspected cointegration relationship is due to a specific economic theory that can be translated into exact parameter restrictions. In such situations, posterior odds ratio tests can be constructed for the restricted parameter space (cointegration) versus the unrestricted parameter space (no cointegration). In fact, Koop includes three hypotheses in his empirical application: stock prices and dividends follow random walks with drift and are cointegrated, stock prices and dividends follow random walks with drift but are not cointegrated, and stock prices and dividends do not follow random walks with drift nor are they cointegrated.

Dorfman (1995) takes a much more general approach, which is better suited to higher-dimensional time series and cases where the parameter values of the hypothesized cointegrating vector are not known or implied by a well-formed economic theory. His method allows for a broader definition of cointegration and focuses the odds ratio test on the posterior support for various numbers of nonstationary roots in the multivariate time series being modeled and the underlying set of univariate time series created by separating the series being studied. In fact, Dorfman places a prior distribution over the number of nonstationary roots in a series (univariate or multivariate), rather than directly on the structural parameters of the VAR models he uses. Dorfman (1995) also allows for uncertainty concerning the lag length of the VAR model. The method is very heavily reliant on numerical methods, specifically importance sampling. The approach results in a numerical derivation of the full posterior distribution of the number of nonstationary roots for the series, which then allows calculation of the posterior probability of cointegration (and even of specific orders of cointegration). Between these two approaches, a Bayesian can investigate such hypotheses as market integration, purchasing power parity, efficient markets in a multiasset framework, along with many others.

DeJong (1992) takes a path somewhere in the middle of these two approaches and investigates the importance of allowing for trend-stationarity as an alternative hypothesis when testing for cointegration. He examines some previously used data sets and finds that placing any prior support on trend-stationarity severely decreases posterior support for cointegration.

Some Applications to the Extended Nelson-Plosser Data

To demonstrate the application of Bayesian numerical methods to cointegration testing, three simple examples were performed using the Dorfman (1995) approach to Bayesian cointegration testing. The data used are all from the extended Nelson-Plosser data set of annual observations that began in years from 1860 to 1909, depending on the series, and end in 1988. The three groups of series chosen for this chapter are: money and real GNP; employment, industrial production, and real wages; and the CPI and GNP deflator. The steps to the three cointegration tests and their results will be presented in sequence. It is worth noting that each test was conducted in less than 15 minutes from specification of the priors to final computation of the posterior odds ratio.

Money and Real GNP

To begin, consider the bivariate series of money and real GNP, both in logs. The data are annual from 1909 to 1988, so there are 80 observations. These series might be cointegrated if the money supply is growing in a constant relationship to the rate of real GNP growth (or vice versa), akin to what Milton Friedman suggested the Federal Reserve Bank should do. However, there is no strong macroeconomic theory or evidence to suggest that such a rule does hold; cointegration between these series is plausible, but not certain.

To test for cointegration, the distribution of the number of nonstationary roots in each individual series and in the bivariate model must be constructed. Model specification uncertainty was allowed within the class of AR(p) and VAR(p) models: models with between 1 and 5 lags were allowed. The prior on the lag lengths places discrete mass of 0.1, 0.2, 0.35, 0.25, 0.1 on the five lag lengths in ascending order. The prior on the number of nonstationary roots was specified as a truncated Poisson distribution with scale parameter equal to 1.25 for the bivariate model and 0.90 for the univariate series. Such a prior places discrete prior weights of 0.305, 0.381, and 0.238 on 0, 1, and 2 nonstationary roots in a bivariate model and weights of 0.464, 0.418, and 0.103 for 0, 1, and 2 nonstationary roots in a univariate model. An inverted Wishart prior is placed on the time series model's error covariance matrix. Finally, these priors were restricted so that positive support was only given to models with structural model parameters less than 2 in absolute values. This makes the priors proper and avoids the risk of arbitrary scaling constants biasing the posterior odds ratio.

The likelihood function was specified by assuming that the errors of the AR and VAR models are normally distributed. Importance sampling with antithetic replication is used to numerically approximate the distribution of the number of nonstationary roots, with the substitute density being the product of the inverted Wishart prior on the error covariance matrix and the likelihood function. Draws are made conditional on each of the five lag lengths (500 antithetic pairs for each specification) and then the lag length uncertainty is integrated out (by averaging across) to arrive at the marginal posterior distribution of the number of nonstationary roots. The three marginal posterior distributions for the number of nonstationary roots (in the bivariate series and in each of the univariate series) are then used to compute the posterior probability of cointegration, which is simply the probability that the bivariate series has fewer nonstationary roots than the two univariate models together. This posterior probability in favor of cointegration can then be used to compute the posterior odds ratio in favor of cointegration relative to the hypothesis of no cointegration. Thus, the final posterior odds ratio for this example is based on 15,000 draws ($3 \times 5 \times 1000$). Further details of the procedure can be found in Dorfman (1995). The results of the cointegration test are not very informative between the two competing hypotheses, with posterior support split fairly evenly between cointegration and no cointegration. This is not too surprising as a story could be told using economic theory to support either view. Other useful results can be gleaned from the empirical exercise, however. The marginal posterior distribution of the lag length for the real GNP series finds the most support for 4 lags, with the posterior mass support at each lag length being 0.033, 0.102, 0.295, 0.344, and 0.226, respectively. Many economists model real GNP as an AR(3) to allow for a trend and a cycle, but this example finds substantial evidence of longer lag lengths being necessary to properly model real GNP.

Finally, the marginal posterior distributions of the number of nonstationary roots is derived as an intermediary result. These results show that the posterior distributions are considerably left-shifted (toward fewer nonstationary roots) relative to the prior distributions, with essentially no posterior support for any of the models being integrated of order 2 (two nonstationary roots). In fact, the posterior evidence is very evenly divided over whether the real GNP series even has a single nonstationary root, with the posterior odds ratio being only barely greater than 1. This finding is relatively consistent with those of Phillips (1991) and DeJong and Whiteman (1991). Table 4 presents a summary of the results of this test, along with the two other cointegration tests; Table 5 shows the marginal posterior distribution of the number of nonstationary roots in the bivariate and two univariate models from this example.

The trivariate example of employment, industrial production, and real wages was chosen with the prior belief that these three series are unlikely to be cointegrated. A theory could be constructed that real wages should only increase in relation to productivity gains which would be increases in industrial production above increases in employment. However, many other variables, which are excluded from this model, would have to be held constant, and some simplifying assumptions about production and quality would have to be made. The prior distributions are

TABLE 4.   Cointegration Test Results for Nelson-Plosser Data

	M-GNP	E-IP-RW	CPI-GNPd
Prior probability of cointegration	0.4728	0.4675	0.4728
Posterior probability of cointegration	0.4441	0.3180	0.8366
Posterior odds ratio	0.7990	0.4662	5.1210
Posterior odds ratio under equal prior odds	0.8909	0.5311	5.7104
Total draws used in numerical approximation	15,000	20,000	15,000
Draws receiving zero prior support	539	967	559

the same as for the money-GNP example. The posterior results show much less support for cointegration for these series, as expected, and are shown in Table 4.

The third example, with the CPI and the GNP deflator, was chosen to prove that the testing procedure is sound. Clearly, these two price indices, which measure such similar baskets of goods, would be expected to be cointegrated. The prior distribution was again the same as for the money-GNP example; 15,000 total draws are used to derive the posterior distribution. As anticipated, these empirical results strongly support cointegration between these two series. The posterior probability of cointegration is 0.8366, making the posterior odds ratio in favor of cointegration equal to 5.121. Thus, the applications with the Nelson-Plosser data have shown that the Bayesian cointegration test used here can produce a split verdict or strong posterior evidence either in favor of or against the hypothesis of cointegration.

TABLE 5.   Distribution of the Number of Nonstationary Roots, Money–Real GNP Example

Number of roots	M-GNP	M	GNP
0	0.310 (0.291)	0.421 (0.426)	0.498 (0.426)
1	0.684 (0.364)	0.579 (0.384)	0.502 (0.384)
2	0.005 (0.228)	0.000 (0.151)	0.000 (0.151)
3	0.001 (0.084)	0.000 (0.035)	0.000 (0.035)

The top number of each pair is the posterior probability, the number underneath (in parentheses) is the prior probability for that number of nonstationary roots.

The most important point of these applications is that having worked out the underlying theory and placed the focus on the number of nonstationary roots, such a Bayesian cointegration test can be performed easily in 10 to 15 minutes, including computing time (about 2 minutes on a 120 MHz Pentium computer). One only needs to specify a few simple prior distributions and to have a computer program that can perform importance sampling with antithetic replication. As one's numerical Bayesian infrastructure of computer code gets built up, the cost of any single Bayesian application quickly decreases to be equivalent to that of sampling theory applications, but the payoff from the Bayesian application is greater due to the richer empirical results that are generated along with simple test statistics such as posterior odds ratios.

References

DeJong, D. N. (1992). "Co-integration and trend-stationarity in macroeconomic time series: Evidence from the likelihood function." *Journal of Econometrics* 52, 347–370.

DeJong, D. N., J. C. Nankervis, N. E. Savin, and C. H. Whiteman (1992). "The power problems of unit root tests in time series with autoregressive errors." *Journal of Econometrics* 53, 323–343.

DeJong, D. N., and C. H. Whiteman (1991). "Reconsidering 'Trends and random walks in macroeconomic time series'." *Journal of Monetary Economics* 28, 221–254.

Dickey, D. A., and W. A. Fuller (1979). "Distribution of the estimators for autoregressive time series with a unit root." *Journal of the American Statistical Association* 74, 427–431.

Dickey, D. A., and S. G. Pantula (1987). "Determining the order of differencing in autoregressive processes." *Journal of Business and Economic Statistics* 5, 455–461.

Dorfman, J. H. (1993). "Bayesian efficiency test for commodity futures markets." *American Journal of Agricultural Economics* 75, 1206–1210.

Dorfman, J. H. (1995). "A numerical Bayesian tests for cointegration of AR processes." *Journal of Econometrics* 66, 289–324.

Geweke, J. (1988b). "The secular and cyclical behavior of real GDP in 19 OECD countries, 1957–1983." *Journal of Business and Economic Statistics* 6, 479–486.

Koop, G. (1991). "Cointegration tests in present value relationships: A Bayesian look at the bivariate properties of stock prices and dividends." *Journal of Econometrics* 49, 105–139.

Koop, G. (1992). "Objective Bayesian unit root tests." *Journal of Applied Econometrics* 7, 65–82.

Phillips, P. C. B. (1991). "To criticize the critics: An objective Bayesian analysis of stochastic trends." *Journal of Applied Econometrics* 6, 333–364.

Phillips, P. C. B., and P. Perron (1988). "Testing for a unit root in time series regression." *Biometrika* 75, 335–346.

Schmidt, P., and P. C. B. Phillips (1992). "LM tests for a unit root in the presence of deterministic trends." *Oxford Bulletin of Economics and Statistics* 54, 257–287.

Schotman, P., and H. K. van Dijk (1991a). "A Bayesian analysis of the unit root in real exchange rates." *Journal of Econometrics* 49, 195–238.

Schotman, P., and H. K. van Dijk (1991b). "On Bayesian routes to unit roots." *Journal of Applied Econometrics* 6, 387–401.

Sims, C. (1988). "Bayesian skepticism on unit root econometrics." *Journal of Economic Dynamics and Control* 12, 463–474.

7

Model Specification Uncertainty

For researchers concerned about model specification uncertainty and the potential of biased coefficient estimates due to either a misspecified model or pretesting of the model specification, a clear solution exists by taking a Bayesian approach to model specification. It is possible in a Bayesian framework to treat the model specification as another unknown parameter to be handled in the same manner as regression coefficients or variance parameters. Model specification parameters will be either integer-valued (such as lag length) or 0/1 variables (whether or not to include a variable). This makes the prior distributions over these parameters discrete mass functions with the prior support distributed across the set of possible model specifications considered. Often the prior distribution over model specification is independent of the prior over the regression parameters (coefficients and variance terms) and can be specified separately, with the two prior distributions multiplied together to form the joint prior distribution.

Allowing multiple-model specifications to receive prior support provides two benefits to the researcher. First, one can derive the posterior support for each of the models in the specified set, thereby gaining insight into the model specification(s) that is (are) most consistent with both the researcher's prior beliefs and the information in the data. Second, one can use the variety of models specified to add robustness to the empirical results. Constructing marginal posterior distributions for parameters (or functions of parameters) by integrating out the model specification uncertainty results in parameter estimates that are highly robust to the effects of any potential model (mis)specification bias. Given the continual presence of model specification uncertainty in economics (what variables to include, what functional form to choose conditional on the variables included), the ability to produce robust parameter estimates is one of the biggest advantages gained from taking a Bayesian approach. Three specific types of model specification uncertainty will be discussed in detail in this chapter: lag length choice in time series analysis, functional form specification for structural models, and the combination of results from different model specifications when possessing only partial information about the individual models. These three types of model specification uncertainty cover the vast majority of those encountered in econometric applications. For example, uncertainty over variable inclusion is identical to functional form questions.

Allowing Lag Length to Be Uncertain

In time series analysis there are virtually no applications in which the appropriate lag length (the number of lagged values of the dependent variables to include on the right-hand side) is known with certainty. Often the range of lag lengths considered is, in fact, quite large, sometimes spanning 12 or 18 different values. Sampling theory statistics has developed estimators for all common time series models that are efficient conditional on the lag lengths (or lengths for ARMA, ARCH, or GARCH models); however, the theory for estimating the lag length is not as well developed. A variety of tests have been proposed for the selection of time series model specification, including the final prediction error (FPE), Akaike information criterion (AIC) and its variations, and the Schwarz criterion (SC), which has a Bayesian derivation.

Rather than attempt to select a single lag length or model specification for a time series model, one can take an approach in which the model specification is treated as a set of parameters. These parameters will be few in number (usually 1 to 3 parameters) and almost always integer-valued. This makes it possible to specify a discrete prior over the set of model specification parameters considered and makes the computation of marginal posterior distributions easier.

Take as an example a univariate ARMA model for a variable y_t that has a zero-mean (for convenience),

$$y_t = \sum_{i=1}^{p} \rho_i y_{t-i} + \varepsilon_t + \sum_{j=1}^{q} \theta_j \varepsilon_{t-j} \tag{7.1}$$

where the ε_t are zero-meaned iid error terms. The two model specification parameters are (p, q), the lag lengths of the autoregressive and moving average terms, respectively. Both p and q are restricted to be non-negative and integer-valued.

Assuming little prior knowledge of the autoregressive and moving average parameter values, a prior distribution for these parameters could be specified as (1) the product of truncated uniform priors with range $(-c, c)$ on each ρ_i and θ_j where c is on the order of 1, 1.5, or 2; (2) a product of independent normal distributions on each ρ_i and θ_j with mean 0 and variance of 1 (or some similar number); or (3) a prior distribution over the roots implied by the ρ_is as discussed in Chapter 6 and some separate prior distribution over the θ_js because they do not influence the long-run dynamics and to specify a prior on only the roots would imply a flat and improper prior over the θ_js. The remaining task is to specify a prior distribution for the model specification parameters.

Because the model specification parameters are integer-valued and restricted to be non-negative, a discrete prior is needed. The first step is to choose the maximum values of p and q that will receive prior support; without such truncation, it will be impossible to compute the posterior distribution of the parameters because an infinite number of models would have to be sampled from in constructing the posterior. This choice is the researcher's, but guidance can be provided based on the observation frequency of the data. Assuming that we are dealing with an economic time series, possible maximum values for the autoregressive lag length are 5 or 7 for annual data, 9 for quarterly data, 18 or 24 for monthly data, 13 or 26 for weekly data, 6 or 11 for daily (five per week) data, and the equivalent of either one or two days plus one for hourly data. Moving average maximum lag lengths can be chosen to match the limit on the autoregressive terms or may be shorter in cases where the maximum autoregressive lag length is particularly long. Note that pure autoregressive or moving average models are simply ARMA models with one of the maximum lag lengths given prior support being equal to zero.

The next step in specifying a prior distribution for the model specification parameters is to assign a prior support mass to each of the possible model specifications included by the researcher's choice of maximum lag lengths. I have used triangular distributions in past applications, placing the prior mode on the central value of the lag lengths receiving support with linearly declining prior weights on each lag length to both sides of the central one. Another prior distribution that I have found appropriate for lag length parameters is a Poisson distribution that has been truncated and renormalized.

Having fully specified prior distributions for the lag length parameters and the structural parameters of the time series model, one can now derive the posterior distribution of any parameters of interest as soon as the likelihood function is specified. Assuming that the researcher is relying on some type of random sampling method for building a numerical approximation to the posterior distribution of some parameter or function of interest, denoted by η, the standard methods presented earlier need only be modified slightly to account for the lag length uncertainty.

To integrate out the lag length parameter and derive the marginal posterior distribution of η, one constructs a posterior distribution for η conditional on each of the values of the lag length parameter(s) that received positive prior support. For each of these saved draws, save the importance weight if using importance sampling and the likelihood function value if using Monte Carlo sampling, Gibbs sampling, or another algorithm that provides draws from the conditional posterior. Assign a new weight for each draw equal to the prior support for that lag length (model specification) multiplied by either the likelihood function value or the importance weight, depending on the sampling method. Combine the draws from the entire set of conditional posteriors, using a weighted average formula equivalent to importance sampling applications with the weights just described that account for the prior support given to the different lag lengths. Remember that any nonconstant prior distributions on other parameters must also be accounted for in the formula. This approach results in the marginal posterior distribution of η and can be used to compute the marginal posterior mean, median, or any percentiles of interest.

To construct the marginal posterior distribution of the lag lengths, one of two approaches is used depending on whether Monte Carlo or importance sampling is being used to generate random draws. With random draws from the joint posterior distribution of all the other parameters conditional on a particular lag length (generated by either Monte Carlo or Gibbs sampling), save the likelihood function value associated with each draw. If the prior weights assigned to each lag length are denoted by ω_j, $j = 1$, p_{max}, where p_{max} is the maximum lag length given positive prior support, and the likelihood function value is denoted by $p(y|\gamma, p)$

where γ represents a vector of all other parameters (the ρ_i, θ_i, and σ^2 parameters from (7.1)) and p is the lag length parameter (possibly vector-valued), then the marginal posterior probability of a particular lag length is given by

$$p(p = j|y) = \frac{\omega_j \sum_{i=1}^{B} p(y|\gamma_j^{(i)}, p = j)^2}{\sum_{j=1}^{p_{max}} \omega_j \sum_{i=1}^{B} p(y|\gamma_j^{(i)}, p = j)^2} \tag{7.2}$$

where the subscript j on the parameter vector γ makes clear that the size of the parameter vector is conditional on the lag length and the B draws made for each model specification are independent of the draws made for the other model specifications. These marginal posterior probabilities constitute the discrete points of the full marginal posterior distribution of the lag length parameter. The marginal posterior mean of the lag length parameter can then be computed as

$$E(p|y) = \sum_{j=1}^{p_{max}} j p(p = j|y). \tag{7.3}$$

If the researcher wants to choose a single lag length specification, the posterior mean is optimal under quadratic loss. Alternatively, nonquadratic loss functions can be used, making the optimal posterior point estimator something other than the posterior mean. For example, Dorfman and Havenner (1992) use a loss function on lag length of the form

$$L(\hat{p}, p) = (\hat{p} - p)^2 D(\hat{p} - p) + c(\hat{p} - p)^2[1 - D(\hat{p} - p)] \tag{7.4}$$

where $D()$ is an indicator function whose value equals 1 when the argument is positive and 0 otherwise and c is a scalar greater than 1. This loss function is an asymmetric quadratic with greater loss for models specified to have lag lengths smaller than the true value. The rationale for such a loss function is that overspecification leads to some sampling error but in large samples causes few problems, while underspecifying the model leads to biased posterior results for other parameters that cannot be overcome by large samples.

In cases where importance sampling is used to generate the random draws, the formula for marginal posterior probabilities of the lag lengths in equation (7.2) must be slightly modified to accommodate the importance weights. If the prior distribution on the model parameters γ conditional on the lag length $p = j$ is denoted by $p(\gamma_j)$, the conditional likelihood function is still represented by $p(y|\gamma_j, p = j)$, and the substitute density used to generate draws for lag length $p = j$ is denoted by $f(\gamma_j, j)$, the modified equation for deriving a marginal posterior probability of lag length j is

$$p(p = j|y) = \frac{\omega_j \sum_{i=1}^{B} p(\gamma_j^{(i)}) p(y|\gamma_j^{(i)}, p = j)^2 / f(\gamma_j^{(i)}, j)}{\sum_{j=1}^{p_{max}} \omega_j \sum_{i=1}^{B} p(\gamma_j^{(i)}) p(y|\gamma_j^{(i)}, p = j)^2 / f(\gamma_j^{(i)}, j)} \tag{7.5}$$

Given these marginal posterior probabilities, marginal posterior means, median, percentiles, or optimal posterior point estimates can be computed or selected, as in the case of Monte Carlo sampling outlined previously. Dorfman (1995) integrates out lag length uncertainty in VAR and AR models to derive marginal posterior distributions of the number of nonstationary roots in exchange rate series that are then used to calculate the posterior probability of cointegration among sets of exchange rates.

Allowing Right-Hand-Side Variable Choice to Be Uncertain

The application of Bayesian methodology to the area of model specification uncertainty in structural econometric models is best exemplified by Poirier (1991). In this paper, Poirier analyzed a variety of macroeconomic questions by entertaining 147 different structural models created by imposing differing theory-based restrictions on various model parameters. Examples include money neutrality and sticky prices. Poirier placed discrete prior mass support on each of these 147 models, continuous prior distributions on the model parameters, and then used Monte Carlo sampling to derive the posterior support for various hypotheses after integrating out the model specification (macroeconomic theory) uncertainty. Given the confusion in macroeconomics over which theories are most appropriate, this is a particularly good application for allowing for a wide variety of possible model specifications.

From a technical standpoint, the treatment of a group of models with different sets of exogenous variables (due to theory restrictions or otherwise) or models with different restrictions on certain parameters is no different from the methodologies presented earlier for treating lag length as uncertain. The only real difference is that often there is no obvious coherent ordering to the set of models given prior support, thus making it somewhat more difficult to assign prior weights to the models. Posterior distributions for parameters, structural or not, and posterior probabilities for each model can be derived exactly as in the preceding lag length example.

Poirier (1991) tried to answer questions about which macroeconomic model structures are best supported by the posterior predictive densities of output and money using a data set that includes 47 countries. By considering seven dynamic specifications for money (unit root, linear trend, etc.), seven dynamic specifications for output, and three specifications concerning the neutrality of money in the output equation, Poirier created a set of 147 models that receive positive prior support distributed in equal measure across the set. Two different prior distributions were constructed for the model parameters, providing some sensitivity analysis in this dimension. Using 100 antithetic replications in a Monte Carlo sampling algorithm for each specification, Poirier computed the posterior support for each of the 147 model specifications for all 47 countries. Thus, although he found that 100 draws was enough to achieve numerical convergence for each model, Poirier's application still includes a total of 690,900 draws; this makes clear the huge advantages that can be gained from antithetic replication. Without it, the computational burden of such a study might have been insurmountable.

Poirier (1991) found that the posterior results differed little across the two prior distributions. Calculation of marginal posterior probabilities for the seven money equation specifications found one posterior probability of 0.907 for one specification in Ecuador and many other cases of specifications receiving more than 50 percent posterior support. The results on output dynamics are similar, varying fairly widely by country but often strongly identifying a particular specification as preferred *a posteriori*. The marginal posterior probabilities of the three money neutrality specifications also found each specification strongly favored for at least one county (for example, U.S. output is not neutral with respect to money shocks while Korea clearly is); in general, money neutrality receives the most posterior support across the 47 countries. Finally, the *a posteriori* most likely model received an average of 0.176 support across the 47 countries, equivalent to 1/6 of the total probability, which is impressive given that each model starts with only 1/147 prior support. The maximum posterior support given to a single model is 0.669, a clear indication that such investigations can produce useful insights into proper econometric model specification and that the prior distributions used by Poirier allowed the data to tell their story.

Another example of treating model specification uncertainty in a Bayesian framework can be found in LeSage (1993), where measures of spatial heterogeneity in regional tobacco production are estimated after integrating out uncertainty over the presence of structural shifts in intercepts and/or slopes and the possibility of outliers. To further increase robustness, results were presented for models with two sets of explanatory variables. Posterior odds ratio tests showed that spatial linkages received support in several states, but in Tennessee there was absolutely no posterior support for such linkages to parameters from other states.

The one caution to applying Bayesian techniques to the area of model specification uncertainty regards the choice of the dependent variables. Because all the posterior inferences rely on optimal combinations of prior information and information from the data that is measured (or viewed) through a likelihood function, the likelihood function's scale must be constant across all the models that are combined when deriving marginal posterior distributions (e.g., ones with the model specification uncertainty integrated out). If two models were given prior support, one with GNP as the dependent variable and the second with ln(GNP), the likelihood functions will be in two different metrics. Combining the two conditional posterior distributions for some parameter η from these two models as outlined earlier would result in incorrect posterior inference on η. This is because the different scales of the two likelihood functions would most likely lead to one model's posterior distribution dominating the other's simply because the magnitude of one model's likelihood is likely to be (arbitrarily) greater than the other's. Thus, it is best to ensure that the dependent variable specification is held constant across all models considered, even if that means that the right-hand side of some econometric models becomes nonlinear.

Model Specification Uncertainty with Partial Information

In the preceding examples, model specification uncertainty was integrated out by constructing a posterior distribution for some parameter of interest η that is a weighted average of a set of posterior distributions for η that were conditional on each model specification receiving positive prior support. Such an approach has only become manageable for ηs that are not simple structural parameters with the advent of numerical integration methods. The numerical approach to this problem, essentially due to Poirier (1991), can only be used to its fullest when the full conditional posterior distributions of η can be generated, meaning that full information consisting of both prior distributions and likelihood function specification must be possessed for each model specification. In some applications, a researcher might have a set of estimators for η generated from different models but no further information.

The most common example of this sort of partial information occurs in the realm of forecasting. A researcher may have a collection of forecasts from a variety of sources but no additional information about the distribution of each individual forecast (in fact, there is often no precise information on the forecasting methods that were used). In such situations, the techniques used by Poirier (1991) are not directly applicable. Instead, a discrete marginal posterior distribution must be constructed based on posterior weights of each source's estimator that are derived from researcher-chosen prior weights that are updated by a performance measure (likelihood function) that is suitable to the particular application.

Imagine a situation in which one wants to predict a time series variable y_t one step ahead. Four forecasts are available to the researcher, denoted by $f_{it}, i = 1, 2, 3, 4$, and the researcher can acquire a past series of forecasts from all four of the sources along with the associated actual values of y_t. Assume that the researcher has no prior information on the performance of the four forecasts; thus, it would be natural to choose equal prior weights on the four forecasts. Denote these prior weights by $\omega_i = 0.25, i = 1, 2, 3, 4$. Next, the performance measure or likelihood function that will be used to map from prior weights to posterior weights must be specified. For this demonstration, let the performance measure be a standard joint normal distribution for the past forecast errors $e_{it} = y_t - f_{it}$,

$$p(e_i) = (2\pi s_i^2)^{-n/2} \exp\left(-\frac{1}{2} \sum_{j=1}^{n} \frac{e_{it}^2}{s_i^2}\right) \tag{7.6}$$

where s_i^2 is the sample variance of the forecast errors and n is the number of past observations on each forecast source. Given the performance measure/likelihood function, the researcher can construct posterior weights for each forecast source. These posterior weights can be denoted by Ω_i, and they are solved for by the formula

$$\Omega_i = \omega_i p(e_i) \bigg/ \sum_{i=1}^{4} \omega_i p(e_i). \tag{7.7}$$

These posterior weights can then be used to compute a Bayesian composite forecast that combines the four individual forecasts using a weighted average with the posterior weights. For each new time period, the Ω_i can be updated as a new observation becomes available for plugging into equations (7.6) and (7.7). Such posterior-weighted composite forecasts are used in Dorfman and Havenner (1992).

The posterior weights can also be used in combination with a loss function to choose a single forecast source as the best forecast for a particular time period, although it is unclear why a Bayesian would want to select a single forecast. If the sample of estimators being combined in such a partial information framework is large enough, one can also compute the posterior median of the estimators using the posterior weights. Such medians are computed for dynamic multipliers in Dorfman and Lastrapes (1996).

More applications of model specification uncertainty can be found in the chapter on forecasting, as this has been a common arena for Bayesian researchers to integrate across multiple possible models in hopes of achieving better forecasting efficiency. As a final note on this topic, the reader is encouraged to read Poirier's (1988) article on econometric model building. If this chapter has not convinced you of the benefits of a Bayesian approach to the model specification uncertainty that is inherent in econometrics, Poirier's elegant discourse on the topic, without dwelling on any of the technical details involved, will convince even the skeptic that all good econometricians should be Bayesian ones.

References

Dorfman, J. H. (1995). "A numerical Bayesian test for cointegration of AR processes." *Journal of Econometrics* 66, 289–324.

Dorfman J. H., and A. M. Havenner (1992). "A Bayesian approach to state space multivariate time series modeling." *Journal of Econometrics* 52, 315–346.

Dorfman, J. H., and W. D. Lastrapes (1996). "The dynamic responses of crop and livestock prices to money-supply shocks: A Bayesian analysis using long-run identifying restrictions." *American Journal of Agricultural Economics* 78, 530–541.

LeSage, J. P. (1993). "Spatial modeling of agricultural markets." *American Journal of Agricultural Economics* 75, 1211–1216.

Poirier, D. J. (1988). "Frequentist and subjectivist perspectives on the problems of model building in economics." *Journal of Economic Perspectives* 2, 121–144.

Poirier, D. J. (1991). "A Bayesian view of nominal money and real output through a new classical macroeconomic window." *Journal Business & Economics Statistics* 9, 125–148.

8

Forecasting

Bayesian approaches offer numerous advantages in the area of forecasting. As mentioned in Chapter 7, forecasts can be made more robust by using a Bayesian approach to form a composite forecast from a set of different forecasting models, thus integrating out model specification uncertainty. Bayesian methodology, with its emphasis on the predictive density of future values (the marginal posterior distribution of y_{t+1} at time t) is extremely well-suited to turning point forecasting as the probabilities of the series moving up or down are easy to calculate. The Bayesian approach is also easily adaptable to multiple-step-ahead forecasting as the uncertainty about intervening periods' values can be accounted for by integrating them out. Basic methodologies applicable to forecasting economic time series will be presented in this chapter along with discussion of a number of good Bayesian forecasting applications.

Basic Forecasting Methods

The basic Bayesian approach to forecasting an economic time series is focused on the derivation of the predictive density. That is, when forecasting y_{t+1} at time t, the predictive density is the marginal posterior distribution of y_{t+1} conditional on all observable variables that are used to help forecast y_{t+1}. While the distribution is conditional on observable data, it is marginal with respect to any unknown parameters. A point forecast can then be made by specifying a loss function for forecasting errors and choosing the point forecast that minimizes the expected loss where the expectation is taken with respect to the predictive density.

For concreteness, take an AR(3) model, a commonly used forecasting model (cf. Zellner and Hong, 1989), and assume that the economic time series variable in question is trend-stationary:

$$y_t = \mu + \delta t + \rho_1 y_{t-1} + \rho_2 y_{t-2} + \rho_3 y_{t-3} + e_t. \tag{8.1}$$

The error term e_t is assumed to be white noise with variance σ^2. Define $\theta = (\mu, \delta, \rho_1, \rho_2, \rho_3, \sigma^2)$ and Y_t as all observations on y_t from time period t backward to the start of the data set. The predictive density for a one-step-ahead forecast is

$$p(y_{t+1}|y_t, y_{t-1}, y_{t-2}) = \int p(y_{t+1}|\theta, Y_t)p(\theta|Y_t)d\theta, \tag{8.2}$$

where $p(\theta|Y_t)$ is the posterior distribution of the unobservable model parameters. Denote a forecast of y_{t+1} by f_{t+1}, and specify a loss function for forecast errors $L(y_{t+1}, f_{t+1})$. The optimal Bayesian posterior point forecast relative to that loss function is then given by

$$f_{t+1} = \operatorname{argmin} \int L(y_{t+1}, f_{t+1})p(y_{t+1}|y_t, y_{t-1}, y_{t-2})dy_{t+1}. \tag{8.3}$$

That is, the point estimate is chosen to minimize the expected loss of any forecast error. Specification of the loss function is up to the researcher, but a variety of loss functions have been used in the literature. A standard quadratic loss function of the form $(y_{t+1} - f_{t+1})^2$ will yield a forecast equal to the mean of the predictive density; an absolute loss function of the form $|y_{t+1} - f_{t+1}|$ yields a point forecast equal to the median of the predictive density.

More complex loss functions can be constructed to produce different optimal point forecasts. Zellner and Hong (1989) used a two-term quadratic loss function to produce shrinkage forecasts of international growth rates; that is, point forecasts that are pushed toward the mean forecast for the entire group of countries whose growth rates are being predicted. Defining y_{it} as the ith country's growth rate at time t, their loss function can be expressed in the notation used here as

$$L(y_{it+1}, f_{it+1}) = (y_{it+1} - f_{it+1})^2 + c\left(\frac{1}{n}\sum_{i=1}^{n}y_{it+1} - f_{it+1}\right)^2, \tag{8.4}$$

where n is the number of countries included in the forecasting exercise. Such a loss function penalizes errors in forecasting the growth rate of a particular economy and predictions of any country vastly differing in performance from the others being studied. If one believes the countries' economies to be linked (through international trade), such shrinkage toward a group mean protects the forecaster from making forecast errors due to the performance of a single country's forecasting model. The amount of shrinkage (or pooling) is controlled by the loss function parameter c, which is selected by the researcher; the larger c, the more tightly bunched the forecasts will become.

In some situations, such as forecasting changes in asset values, an asymmetric loss function is advantageous. Generally, professional forecasters prefer to overpredict downward movement in asset prices and underpredict price increases because over many years they have found that investors prefer to sell an asset too soon rather than too late. Thus, if y_t is a time series on the price of an asset (stock, bond, etc.), a Bayesian stockbroker might use a loss function such as

$$L(y_{t+1}, f_{t+1}) = D(y_{t+1}, f_{t+1})(y_{t+1} - f_{t+1})^2 + c[1 - D(y_{t+1}, f_{t+1})](y_{t+1} - f_{t+1})^2 \tag{8.5}$$

where $D(y_{t+1}, f_{t+1})$ is an indicator function that equals 1 when $f_{t+1} < y_{t+1}$ and 0 when $f_{t+1} \geq y_{t+1}$, and c is a positive scalar greater than 1. As c becomes larger, the penalty for overstating the potential gains from an investment is increased and the minimum expected loss forecast will decrease.

With many time series models, such as the AR(3) model with trend displayed in equation (8.1), if a Jeffreys prior is taken for θ ($p(\theta) \propto 1/\sigma$), the predictive density of y_{t+1} takes the form of a Student-t distribution and the mean of the predictive density is easy to calculate. For more complex loss functions, more complex prior distributions, or more complex forecasting models (particularly models that are nonlinear in the parameters), computation of either the predictive density or the minimum expected loss point forecast can be difficult to do analytically. In such instances, the numerical methods that are the focus of this book are well-suited to performing the necessary calculations. Computing the mean (or median) of a predictive density is a straightforward application of Monte Carlo sampling if it can be drawn from directly or of importance sampling otherwise.

The procedure for developing fully Bayesian multiple-step-ahead forecasts was first presented in a numerical context by Thompson and Miller (1986). Thompson and Miller laid out an early numerical approach to approximating the predictive densities of multiple-step-ahead forecasts for univariate time series models that integrated out parameter uncertainty and the uncertainty associated with the intervening unobservable time series values between the current period and the future time period for which the time series was being forecast. This paper and Monahan's (1983) application to numerical integration in ARMA models were early demonstrations of the benefits that were to be derived from advances in numerical integration.

Finding the minimum expected loss forecast for more complex loss functions generally requires a slightly more complicated approach, with a grid search loop added to the sampling algorithm. When the minimum expected loss forecast is not equal to the mean or median of the predictive density, the point forecast that minimizes the expected loss function is found in two steps. First, generate a set of draws from the predictive density using an appropriate sampling method, probably either importance or Monte Carlo sampling. Next, search over a range of possible values for f_{t+1}, calculating the expected loss associated with each value considered using the set of draws that were generated from the predictive density to numerically approximate the analytical formula shown in the right-hand side of equation (8.3). Save the expected loss for each candidate forecast value and continue, using the same set of draws to evaluate each candidate forecast. When an expected loss value has been computed and saved for each candidate forecast, choose the forecast with minimum expected loss as the optimal Bayesian point forecast.

An easy way to choose candidate values is to use the minimum and maximum values in the set of draws from the predictive density as the extremes of the values considered. Then divide this range into 100 or 1000 candidate forecasts, compute the expected loss of each candidate forecast, and then find the minimum expected loss forecast from within this set of candidates. If this division does not provide enough numerical accuracy for the point forecast, repeat the search procedure on a second, finer mesh centered around the first grid search's estimated minimum expected loss forecast with a much smaller range and smaller increments between candidate forecasts.

A Turning Point Forecasting Application

Zellner and a number of co-authors have written a series of papers on turning point forecasting with a simple, straightforward approach that is easy to implement (Zellner and Hong, 1988; Zellner, Hong, and Gulati, 1990; Zellner, Hong, and Min, 1991). Using an autoregressive model as in equation (8.1) but with the addition of leading indicator variables to improve prediction, they have demonstrated the ability of Bayesian forecasting methods to predict turning points in international economic growth rates with admirable accuracy. The models, termed AR(3)LI models for the leading indicators that are included (all lagged to ensure that they are observable at the time the forecast needs to be made), can be written as

$$y_t = \mu + \rho_1 y_{t-1} + \rho_2 y_{t-2} + \rho_3 y_{t-3} + x_{t-1}\beta + e_t = z_{t-1}\theta + e_t. \qquad (8.6)$$

In equation (8.6), the row vector x_{t-1} contains the relevant lagged leading indicator variables (generally, measures of growth in the money supply, average returns in the country's stock market, and median world stock returns).

If a very diffuse normal-inverse gamma prior is placed on the structural coefficients $\theta = (\mu, \rho_1, \rho_2, \rho_3, \beta)$ and on the error's standard deviation, σ, such as $p(\theta|\sigma) \sim N(0, \sigma^2 I_5 \times 10^6)$ and $p(\sigma) \sim IG(v)$, then the predictive density of the one-step ahead forecast f_{t+1} will be a mean-shifted univariate Student-t distribution so that the quantity

$$(f_{t+1} - z_t\theta)/s_t a_t \sim t(v + t) \qquad (8.7)$$

where $\hat{\theta}$ is the current OLS (ML) estimate of the regression model parameters, s_t is the current OLS estimate of σ, $a_t = 1 + z_t(Z_t'Z_t)^{-1}z_t'$, and Z_t is the current matrix of regressors over the entire sample period up to time t. That is, the quantity on the left-hand side of equation (8.7) is distributed as a Student-t random variable with $(v + t)$ degrees of freedom.

The posterior probability of movement down or up can then be calculated as either the integral of the predictive density from negative infinity to y_t or from y_t to positive infinity, respectively. Given the variables in equation (8.7), either integral can be found by using a standard t-table or software command for calculating tails of the Student-t distribution.

Forecasts of the direction of movement in the time series y_t are then made from these posterior probabilities through the application of a loss function for the forecast errors. Because there are only four possible events in the forecasting sample space, the expected losses of the two possible forecasts are easy to compute. Assume that correct forecasts of movement either up or down have no loss, while incorrect forecasts of movement up cause loss 1 and incorrect forecasts of movement down cause loss c (a positive scalar that can be less than, equal to, or greater than 1). Then denoting the posterior probabilities of movement up and down computed using the distribution of equation (8.7) by P_u and $P_d = 1 - P_u$, the expected losses from forecasting movement up or down are given by

$$EL(\text{predict up}) = P_d = 1 - P_u \qquad (8.8)$$

$$EL(\text{predict down}) = c P_u. \qquad (8.9)$$

Minimization of expected loss implies that a forecast of movement up should be made whenever $P_u > 1/(1 + c)$. The references given at the beginning of this section contain numerous applications of this general method to economic growth rates from 18 OECD countries.

Composite Forecasting

As mentioned in Chapter 7, model specification uncertainty has a natural connection with the area of forecasting. In many applications, researchers are interested in forming a composite forecast using a set of forecasts derived from competing forecasting models. Bayesian methods for forming such composite forecasts are straightforward and easy to implement. Two such methods will be detailed here: first, a standard approach when full information is possessed about all forecasting models considered, and, second, a partial information approach due to LeSage and Magura (1992) that is appropriate when a researcher only has the forecasts from the different models, not full information on the models used to generate the forecasts. A good discussion of Bayesian methods for combining forecasting models can be found in Min and Zellner (1993).

A Full Information Approach

For simplicity, assume we have three competing forecasting models, each worthy of receiving positive prior support for the validity of its forecasts. Denote the models used to generate the forecasts by

$$y_t = x_{it}\beta_{it} + e_{it}, i = 1, 2, 3. \qquad (8.10)$$

In each model, denote the matrices of all observations up to date t by Y_t and X_{it}, and the current one-step-ahead forecast by f_{it}. If forecasting begins with time period $s > \text{cols}(x_{it}), i = 1, 2, 3$, the most basic way to form a composite forecast is with the posterior odds of each model.

Assign each model a positive prior weight ω_i such that the three prior weights sum to 1. The posterior odds of each model for time period s are then given by the normalized products of the prior weights and the likelihood function of each model, $p(Y_s|X_{it}, b_{it}, s_{it})$ where b_{it} and s_{it} are the current estimates of β_{it} and the standard deviation of the error term estimated with the prior distribution, likelihood function, and loss function associated with forecasting model i. Note that the likelihood function is based on the current estimate of the model parameters and evaluated over all observable data, so that period $s + 1$'s likelihood function value will not be equal to period s's multiplied by the probability density of the additional observation on y_{s+1}. Defining the posterior odds of each model at time s by Ω_i, the posterior odds can be written mathematically as

$$\Omega_{is} = \frac{\omega_i \, p_i(Y_t | X_{it}, b_{it}, s_{it})}{\displaystyle\sum_{i=1}^{3} \omega_i \, p_i(Y_t | X_{it}, b_{it}, s_{it})}. \tag{8.11}$$

If the forecasts from each of the three models are denoted by f_{is}, then the posterior odds composite forecast is simply a weighted average of the three component forecasts using the posterior odds as the weights,

$$f_s = \sum_{i=1}^{3} \Omega_{is} f_{is}. \tag{8.12}$$

LeSage and Magura's Partial Information Approach

LeSage and Magura (1992) present a Bayesian approach to combining forecasts when the only information possessed by the researcher is the forecasts generated by the individual forecasting models. The researcher does not need to know the models used to produce the forecasts or even the variables used in the models. The method relies on a set of recursive formulas to update posterior distributions for the parameters of a set of dynamic linear models (West and Harrison, 1989) that are used to model the composite weights. This procedure can be done at a simplified level, or the full LeSage and Magura (1992) approach can be used.

In the simplified form, the dynamic linear model approach to composite forecasting begins by representing the composite forecast f_t as a linear combination of an intercept and the $(m - 1)$ component model forecasts z_{it}. The dynamic linear model (a type of time-varying parameter state-space model) relating the component forecasts to the composite can be expressed in two equations as

$$f_t = z_t \beta_t + v_t \tag{8.13}$$

$$\beta_t = \beta_{t-1} + w_t \tag{8.14}$$

where z_t is a $(1 \times m)$-row vector containing an intercept and the component forecasts z_{it}, the β_t are time-varying composite weights to be estimated, and v_t and w_t are white noise error terms.

Estimation of this model begins with priors on β_0 and the variances of the two error terms. To allow analytical results to be derived, the natural conjugate priors are

$$\beta_0 \sim N(\mu_0, C_0), \; \Phi \sim G(n_0/2, d_0/2), \tag{8.15}$$

where $G()$ represents a gamma distribution, $\Phi = 1/E(v_t^2)$, and the prior on $W_t = E(w_t w_t')$ is covered by the specification of C_0. The current period estimates are then updated from those of the previous period by the recursion formulas:

$$s_{t-1} = d_{t-1}/n_{t-1} \tag{8.16}$$

$$R_t = C_{t-1} + W_t \tag{8.17}$$

$$q_t = z_t R_t z_t' + s_{t-1} \tag{8.18}$$

$$n_t = n_{t-1} + 1 \tag{8.19}$$

$$d_t = d_{t-1} + (f_t - z_t \mu_{t-1})^2 s_{t-1}/q_t \tag{8.20}$$

$$a_t = (z_t R_t)'/q_t \tag{8.21}$$

$$\mu_t = \mu_{t-1} + a_t(f_t - z_t \mu_{t-1}) \tag{8.22}$$

$$C_t = R_t(s_t/q_t) \tag{8.23}$$

So when a forecast is made, the distribution of the composite weights β_t is $\beta_t \sim N(\mu_{t-1}, R_t)$ and the posterior mean forecast of f_t (the optimal point forecast under quadratic loss) is $z_t \mu_{t-1}$. After observing the forecast error $(f_t - z_t \mu_{t-1})$, the new posterior distribution of β_t is $\beta_t \sim N(\mu_t, R_t)$. This update comes too late to use in real-time forecasting of the time series f_t. The reason that μ_{t-1} can be used as the prior mean of β_t is the particular form of the state-transition equation for changes in the composite weights that was assumed in equation (8.14), namely, a random walk. If the state transition equation (8.14) is generalized to allow for a more flexible evolution of the composite weights, there would be a corresponding adjustment in the distribution of β_t when the forecast must be made; see West and Harrison (1989) or Pole, West, and Harrison (1994) for details.

Such a dynamic linear model can be used to form composite forecasts as outlined earlier, with the composite weights β_t evolving through time as the posterior distribution for the weights gets updated after each observed forecast error. The procedure will automatically increase the weights on component models that are more accurate and shrink their variance components in R_t, making those weights become more stable.

More complex dynamic linear models than the one presented here can be estimated, even with analytical distributional updates as outlined earlier. The reader is encouraged to find further details in West and Harrison (1989) and Pole, West, and Harrison (1994). The second reference even comes with a computer disk containing software to implement dynamic linear models of many different varieties, from simple to complex. LeSage and Magura (1992) use a dynamic linear model virtually identical to the one presented earlier as well as four more complex models with parameter shifts at certain breakpoints in the time series. They then use this set of five models to incorporate model specification uncertainty into their composite formation process. At each point in time, they derive probabilities of the last period being best represented by one of the five models and the next period's probability of belonging to each of the five model types, thus resulting in a total of 25 possible posterior distributions based on the combination of potential updates. LeSage and Magura found that this multiprocess mixture model works well in forming composite U.S. GNP forecasts from four leading economic forecasting firms. However, their approach is quite complex, and a researcher would need a strong belief in the necessity of such a flexible structure to justify the additional technological investment above a single standard dynamic linear model.

Composite Qualitative Forecasting

Li and Dorfman (1995, 1996) have proposed a method for forming composite forecasts from a set of individual qualitative forecasts when the person forming the composite has very little information in addition to the set of component forecasts. Their approach is designed specifically for qualitative forecasts and generates composite weights that attempt to measure the relative probability of each component forecast's being correct in the current forecast period. Assume that the researcher has a set of dichotomous (0/1) forecast variables, such as forecasts of movement up or down in an asset price. Denote these forecasts by z_{it}, $i = 1, \ldots, m$.

To implement the Li-Dorfman composite method, the researcher first specifies a set of observable exogenous variables that relate to the accuracy of the component forecasts; denote these variables by x_t. Li and Dorfman (1995, 1996) chose variables related to the current market conditions in the area of economic activity associated with the variable being forecast. For example, if the zs are forecasts of upward or downward movement in the GNP of a country, the variables included in the x_t vector might be lagged GNP growth rate, periods since the last downturn in economic activity, lagged stock market returns (following the research of Zellner and his co-authors), lagged money growth rate, and so on. All variables in x_t must be carefully chosen to ensure that they are observable when the composite forecast is formed.

Using a set of past data on each of the individual forecast models (whose exact form can remain unknown) and the actual occurrences of the dichotomous variable being forecast, a logit model is estimated for each individual forecast model measuring the ability of each component model to accurately forecast the event at issue. The estimation of these logit models can be Bayesian in approach (see Li and Dorfman, (1995, 1996) and Pratt, Raiffa, and Schlaifer (1965) for details of methods that provide Bayesian motivation for estimators that match the familiar maximum likelihood estimators; see Zellner and Rossi (1984) for approaches to estimating logit models under informative priors), but it does not need to be. Then, using the most current observable values for the x_t variables, the probability of each model's current period forecast being correct is predicted using the estimated logit models,

$$P_{it} = \text{probability(model } i\text{'s forecast for period } t \text{ is correct)}$$

$$= \exp(x_t b_{it})/[1 + \exp(x_t b_{it})], \tag{8.24}$$

where b_{it} is the current estimated logit model parameters from model is logit model.

These P_{it} are then normalized so that they sum to 1 over the m individual models; the set of normalized probabilities, p_{it}, are the composite weights in the

Li-Dorfman method. A composite forecast is then generated according to the rule

$$z_t = \begin{cases} 1 & \text{if } \sum_{i=1}^{m} p_{it} z_{it} > 1/(1+c) \\ 0 & \text{otherwise} \end{cases} \qquad (8.25)$$

where c is the loss from incorrectly predicting the outcome represented by 0 relative to the loss from incorrectly predicting the outcome represented by 1. Li and Dorfman re-estimated the logit models after each period's results became observable and then generated new composite weights for the next period's forecasts. They found that the composite weights generated by this method are highly flexible, allowing individual models to be weighted heavily or virtually ignored from period to period. Successful empirical results in applications to employment data and international economic growth rates indicate that this method of forming composite qualitative forecasts has good potential.

References

LeSage, J. P. and M. Magura (1992). "A mixture-model approach to combining forecasts." *Journal of Business & Economic Statistics* 10, 445–452.

Li, D. T., and J. H. Dorfman (1995). "A robust approach to predicting fluctuations in state-level employment growth." *Journal of Regional Science* 35, 471–484.

Li, D. T., and J. H. Dorfman (1996). "Predicting turning points through the integration of multiple models." *Journal of Business & Economic Statistics* 14, 421–428.

Min, C. K., and A. Zellner (1993). "Bayesian and non-Bayesian methods for combining models and forecasts with applications to forecasting international growth rates." *Journal of Econometrics* 56, 89–118.

Monahan, J. F. (1983). "Fully Bayesian analysis of ARMA time series models." *Journal of Econometrics* 21, 307–331.

Pole, A., M. West, and J. Harrison (1994). *Applied Bayesian Forecasting and Time Series Analysis*. New York: Chapman-Hall.

Pratt, J. W., H. Raiffa, and R. Schlaifer (1965). *Introduction to Statistical Decision Theory*, New York: McGraw-Hill Book Company.

Thompson, P. A., and R. B. Miller (1986). "Sampling the future: A Bayesian approach to forecasting from univariate time series models." *Journal of Business & Economic Statistics* 4, 427–436.

West, M., and J. Harrison (1989). *Bayesian Forecasting and Dynamic Models*. New York: Springer-Verlag.

Zellner, A., and C. Hong (1988). "Bayesian methods for forecasting turning points in economic time series: Sensitivity of forecasts to asymmetry of loss structures" *Leading Economic Indicators: New Approaches and Forecasting Records*, K. Lahiri and G. Moore, Eds., Cambridge: Cambridge University Press.

Zellner, A., and C. Hong (1989). "Forecasting international growth rates using Bayesian shrinkage and other procedures." *Journal of Econometrics* 40, 183–202.

Zellner, A., C. Hong, and G. M. Gulati (1990). "Turning points in economic time series, loss structures and Bayesian forecasting," *Bayesian and Likelihood Methods in Statistics and Econometrics: Essays in Honor of George A. Barnard*, S. Geisser, J. Hodges, S. J. Press, and A. Zellner, Eds., Amsterdam: North Holland, pp. 371–393.

Zellner, A., C. Hong, and C. Min (1991). "Forecasting turning points in international output growth rates using Bayesian exponentially weighted autoregression, time-varying parameter, and pooling techniques," *Journal of Econometrics* 49, 275–304.

Zellner, A., and P. E. Rossi (1984). "Bayesian analysis of dichotomous quantal response models." *Journal of Econometrics* 25, 365–393.

9

More Realistic Models Through Numerical Methods

This chapter will provide a brief list of some of the econometric models that are now open to Bayesian estimation and hypothesis testing, that before the recent advances in numerical integration would have been impossible to analyze in a Bayesian framework. The common thread for most of these models is the presence of error terms with distributions that are more complicated to work with analytically than normal iid random variables. Few details of the actual applications are given here, as such models tend to be highly specific to the particular settings and data to which they are applied. Instead, the reader is directed to a published application for the full theory and methods involved. However, this chapter will hopefully serve as both a guidepost to good Bayesian applications with nonstandard models and an advertisement that such complex models are now available for Bayesian inspection.

Frontier Models

Production and cost frontier models have become more common over the last ten years because econometric and mathematical programming methods for analyzing such models have been developed. Such models are applied frequently in studies of technical and allocative efficiency. A common formulation of a production frontier might consist of a regression model such as

$$ln(q_{it}) = \beta_0 + \beta_1 ln(k_{it}) + \beta_2 ln(h_{it}) + \beta_3 ln(s_{it}) + e_{it} + v_i \qquad (9.1)$$

where q is quantity produced, k is capital, h is hours of labor, s represents other inputs, the subscripts i and t represent firms and time periods, respectively, e is a standard zero-meaned white noise error, and v_i is a firm-specific unobservable component representing technical inefficiency (Aigner, Lovell, and Schmidt, 1977). Because the v_i are constrained by definition to be negative, some non-normal distribution must be assumed for the v_i; common choices have been truncated and half-normal distributions. Gibbs sampling would be a natural method for analyzing the joint posterior distribution of the βs and the parameters controlling the distributions of the es and vs. Such an approach would also provide an easy method to calculate the entire posterior distribution of the technical efficiency measures, rather than producing only the point estimates that have been commonplace in the efficiency literature. Bayesian applications of stochastic frontier models can be found in van den Broeck, Koop, Osiewalski, and Steel (1994) and Koop, Osiewalski, and Steel (1997). The latter paper discusses models suitable for estimation by Monte Carlo sampling and models that are best estimated using Gibbs sampling. Full details are provided concerning priors, likelihood, and the implementation of the Gibbs sampling algorithm.

ARCH Models

Financial asset price series have often been found to be poorly represented by ARMA time series models with normally distributed innovations. Instead, much research has shown that the tails of the distributions of price changes are fatter than characterized by normality and that the variance of the innovations may well change over time (cf. Boothe and Glassman, 1987). A variety of models have been proposed to deal with these empirical facts, with one of the more popular recent models being the ARCH or autoregressive with conditional heteroscedasticity model that allows the variance of the innovations (error terms) to follow an autoregressive process (Engle, 1982; Engle and Bollerslev, 1986). Such models are complicated to estimate by sampling theory methods because the time-varying error variances are unobservable and the parameters estimated for the variance's autoregressive process must satisfy certain inequality restrictions that are implied by variances always being positive. Geweke (1989) developed an application to daily stock price data that shows how to estimate an ARCH model in a Bayesian framework. He began with the Jeffreys prior for this model, including the inequality restrictions on the positive prior support for the parameter space, specified the likelihood function, and then derived the posterior distribution of all the parameters using importance sampling. Geweke also detailed the derivation of the predictive density of multiple-step-ahead forecasts from the ARCH model. Geweke's application even included posterior odds ratio tests for model specification uncertainty over both nested and non-nested specifications.

Qualitative Choice Models

The most common qualitative choice models are the probit and logit models that are applied in situations with dichotomous dependent variables. The best source for describing the Bayesian estimation of basic qualitative choice models is Zellner and Rossi (1984). Zellner and Rossi laid out analytical approximations to the posterior distributions of the parameters of both logit and probit models, under diffuse and informative priors. They also present numerical approaches to deriving the exact posterior distributions using importance sampling and show through a Monte Carlo exercise the benefits of numerically approximating the exact distributions of interest.

For situations with multiple choice possibilities, McCulloch and Rossi (1994) present a Gibbs sampling approach to numerically approximate the posterior distribution of the parameters of a multinomial probit (MNP) model. The MNP model is extremely difficult to estimate by maximum likelihood techniques for choice sets of more than three or four choices due to the multidimensional integrals that must be performed in the evaluation of the likelihood function. A few researchers have used approximations such as the method of simulated moments in an attempt to circumvent these difficulties (McFadden, 1989). The Gibbs sampling approach of McCulloch and Rossi (1994) is easier to implement than any of the feasible sampling theory approaches and provides full posterior distributions of the parameters, not just point estimates. Dorfman (1996) applied the McCulloch and Rossi (1994) Gibbs sampling approach to the MNP model to data on technology adoption; this paper includes the formulas for calculating the posterior distributions of the marginal probabilities, the impact on choice probabilities of marginal changes in the explanatory variables.

Koop and Poirier (1994) present an application using a rank-ordered multinomial logit model to study voter preferences. They derived results for several priors, including conjugate priors, and they provide a specification test.

Models with Non-normal Error Distributions

Like the frontier models discussed at the start of this chapter, some econometric applications are not well-suited to the assumption of normally distributed errors, even with corrections such as ARCH effects. Finance data is well-documented to fail tests for normally distributed errors, and many other applications maintain normality without testing for it. There are many probability distributions with more flexibility than the normal distribution that might be better suited to a particular application. With numerical integration, it is often no harder to generate draws from a non-normal distribution, so that making more realistic assumptions about the error distribution may not complicate the analysis.

The main reason for maintaining the assumption of normality in econometrics is hypothesis testing, because least squares estimation does not rely on any distributional assumption; normality is needed to perform the t-tests, F-tests, and χ^2 tests with which we are so familiar. In numerical Bayesian applications, hypothesis testing can be performed as easily on posterior distributions of any form because one is working with an empirical approximation of the posterior distribution and no analytical calculations are necessary. One is only limited by the ability either to generate draws from the posterior distribution or to evaluate the posterior distribution at a point drawn from a suitable substitute density (in the case of importance sampling). Given a good reference on generating random numbers from different probability distributions, these limits still allow for broad latitude in selecting appropriate error distributions. One such reference is Kennedy and Gentle (1984); the reader can choose a personal favorite from several other available works on the topic of statistical computing.

It is worth separate mention that considerable work has been devoted to generating priors and methodologies to allow the Bayesian investigation of models with elliptical errors and with Student-t errors, both classes more flexible than those with normal errors, but still allowing some analytical results. For two examples, see Geweke (1993) and Koop and Steel (1994).

References

Aigner, D., C. A. K. Lovell, and P. Schmidt (1977). "Formulation and estimation of stochastic frontier productions function models." *Journal of Econometrics* 6, 21–37.

Boothe, P., and D. Glassman (1987). "The statistical distribution of exchange rates: empirical evidence and economic implications." *Journal of International Economics* 22, 297–319.

van den Broeck, J., G. Koop, J. Osiewalski, and M. F. J. Steel (1994). "Stochastic frontier models: A Bayesian perspective." *Journal of Econometrics* 61, 273–303.

Dorfman, J. H. (1996). "Modeling multiple adoption decisions in a joint framework." *American Journal of Agricultural Economics* 78, 547–557.

Engle, R. (1982). "Autoregressive conditional heteroscedasticity with estimates of the variance of United Kingdom inflation." *Econometrica* 50, 987–1008.

Engle, R., and T. Bollerslev (1986). "Modeling the persistence of conditional variances." *Econometric Reviews* 5, 1–50.

Geweke, J. (1989). "Exact predictive densities for linear models with ARCH disturbances." *Journal of Econometrics* 40, 63–86.

Geweke, J. (1993). "Bayesian treatment of the independent Student-t linear model." *Journal of Applied Econometrics* 8, S19–S40.

Kennedy, W. J., Jr., and J. E. Gentle (1984). *Statistical Computing*. New York: Marcel Dekker.

Koop, G., J. Osiewalski, and M. F. J. Steel (1997). "Bayesian efficiency analysis through individual effects: Hospital cost frontiers." *Journal of Econometrics* 76, 77–105.

Koop, G., and D. J. Poirier (1994). "Rank-ordered logit models: An empirical analysis of Ontario voter preferences." *Journal of Applied Econometrics* 9, 369–388.

Koop, G., and M. F. J. Steel (1994). "A decision theoretic analysis of the unit root hypothesis using mixtures of general elliptical models." *Journal of Business & Economic Statistics* 12, 93–106.

McCulloch, R., and P. E. Rossi (1994). "An exact likelihood analysis of the multinomial probit model." *Journal of Econometrics* 64, 207–240.

McFadden, D. (1989). "A method of simulated moments for estimation of discrete response models without numerical integration." *Econometrica* 57, 995–1027.

Zellner, A., and P. E. Rossi (1984). "Bayesian analysis of dichotomous quantal response models." *Journal of Econometrics* 25, 365–393.

Part III

Applications to Economic Decision Making

10

Decision Theory Applications

The Bayesian approach to decision theory allows a researcher to find the optimal setting for a set of control variables with respect to the posterior distribution of any unknown random parameters and a specified objective function. In estimation applications, the Bayesian approach was contrasted to the sampling theory statistics, but in the area of decision theory, the Bayesian approach contrasts with the "plug-in" approach. In the plug-in approach to decision problems, the researcher takes estimates of unknown parameters as if they were certain and ignores the effect that parameter uncertainty may have on choosing the optimal set of control variables. The certainty equivalence principle states that such ignorance of parameter uncertainty will lead to the same solution as a Bayesian approach that fully accounts for the parameter uncertainty if and only if the posterior distribution of the parameters is normal and the objective function is linear-quadratic in the unknown parameters. In all other cases, the two approaches will provide different answers for the optimal control variable settings. The Bayesian decision theory approach is also often referred to as decision making under estimation risk to stress the difference between the plug-in approach of assuming that parameters are estimated perfectly and the Bayesian approach, which incorporates the uncertainty created by working with parameter estimates into the solution for an optimal decision.

The basic Bayesian approach to a decision theory problem is simple and straightforward. One chooses the values of the control variables that minimize the expected loss of that decision, where the expectation is taken with respect to the posterior distribution of the unknown parameters (Berger, 1985). Recalling the discussion in Chapter 2, if the unknown parameters involved in the decision are θ with posterior distribution $p(\theta|y, x)$, the objective function to be minimized (or maximized) is $F(y, x|\theta)$, y is a vector of random variables with density $p(y|x, \theta)$, and x is the vector of control variables, the Bayesian decision problem can be written as

$$\text{choose } x = \text{argmin} \iint F(y, x|\theta)p(y|x, \theta)p(\theta|y, x)\, dy\, d\theta. \qquad (10.1)$$

In equation (10.1), the expected loss of choosing a particular (vector) value of the control variable x is evaluated by integrating out uncertainty concerning the unknown parameters θ and any residual uncertainty about the endogenous variables y.

If the value of y is deterministic given x and θ, $p(y|x, \theta)$ will be degenerate and will not affect the calculation of expected loss. The optimal choice of x is then found as the value that minimizes the expected loss (argmin is mathematical shorthand for the value of the argument of minimizing the specified function or functional).

While the preceding discussion above talked about minimizing the expected loss associated with a decision, the common economic criterion of maximizing profit can easily be fit into this paradigm. Define the loss function to be $F(q|\theta) = \pi_{max} - \pi(q|\theta)$, where q is the quantity produced of a good, θ are the unknown parameters, and $\pi(q|\theta)$ is the profit earned from producing and selling quantity q conditional on the parameters θ. The value of π_{max} will be unknown, but because it is a constant, we can define a new (computable) loss function, $F^*(q|\theta) = -\pi(q|\theta)$, which will be minimized by the same value of q as the original loss function $F(q|\theta)$. Minimizing the negative of profit is equivalent to maximizing profit, so using $F^*(q|\theta)$ as a loss function whose expected value is to be minimized will result in a decision that maximizes the expected profit from the decision.

An Application to Land Allocation

Lence and Hayes (1995) performed a very good application of Bayesian decision theory to the problem of land allocation. Imagine a farmer who has available a fixed amount of land that can be planted in any combination of up to four different crops. Assume that the farmer's objective is to choose the acreage allocation that maximizes the expected utility of profit from the crops grown. Lence and Hayes (1995) chose a negative exponential utility function with Arrow-Pratt absolute risk aversion coefficient of ϕ. For the presentation here, the vector of decision (control) variables is denoted by acreage allocations to the four (arbitrarily ordered) crops, $d = (d_1, d_2, d_3, d_4)$, where $d_4 = L - d_1 - d_2 - d_3$ as long as all four crops have sufficiently profitable expected returns to ensure full allocation of the acreage. The per-acre profit of each crop is an unknown random parameter when planting decisions are made; group the four random profit variables into a conformable vector π. The loss function for the minimization problem can therefore be written

$$F(d|\pi) = \exp(-\phi\pi d), \tag{10.2}$$

where there is no negative in front of the exponential function because we are minimizing instead of maximizing. Using two published sets of profits and their variances, Lence and Hayes imagined that the farmer takes a Jeffreys prior for the profits and variance-covariance matrix, resulting in the derivation of a marginal posterior distribution for the per-acre profits in the form of a multivariate Student-t distribution that we can represent here as $p(\pi|x) = MVT(\mu, \Sigma, k)$ where μ is the mean per-acre profits observed by the farmer in the data used to derive the posterior, Σ is the scaling matrix (posterior variance), k is the degrees of freedom, and x is the observed data on profits.

To numerically approximate the optimal acreage allocation, Monte Carlo sampling can be used because draws can be drawn easily from the posterior distribution (a multivariate Student-t form). First, one draws a set of B simulated profits for each of the four crops from the posterior distribution (Lence and Hayes used 25,000 draws). With the draws in hand, the expected loss of a particular acreage allocation can be calculated, beginning with plugging the allocation into the loss function in conjunction with each of the random parameter draws to get a single value from the posterior distribution of the loss function,

$$F(d_j|\pi^{(i)}) = \exp(-\phi\pi^{(i)} d_j). \tag{10.3}$$

In equation (10.2), the subscript j indexes possible acreage allocation choices while the superscripted (i)s denote the draws from the posterior distribution of the random parameters. The expected loss of a given acreage allocation is then calculated by a simple arithmetic average of the B individual loss function values described by equation (10.3):

$$E[F(d_j)] = \sum_{i=1}^{B} \exp(-\phi\pi^{(i)} d_j). \tag{10.4}$$

To find the optimal acreage allocation, one needs to scan the set of potential acreage allocations d_j and choose the one that minimizes the expected loss. The scan over potential allocations can be done in several ways. A gradient-based descent method can be used to actively search for the optimal allocation or a passive grid search method can be used.

Two types of ordered grid searches can be performed. The first is designed to find the optimal acreage allocation in one scan. Assume that the total acreage L can be divided up in units no smaller than one acre (in many real-world applications, the units would be much larger than this). That limit controls the minimum increment of the grid search. With four possible crops, the grid search would consist of a three-level nested loop because the fourth allocation is determined by the first three. The outer loop is the allocation to the first crop, the second loop controls the allocation to the second crop, the third loop will control the third crop. Let the second loop begin at $d_2 = L - d_1$ and let the third loop begin at $d_3 = L - d_1 - d_2$. Begin with a 100 percent allocation to the first crop and calculate the expected loss. The process would produce trial allocations that followed a pattern such as $(L, 0, 0, 0)$, $(L-1, 1, 0, 0), (L-1, 0, 1, 0), (L-1, 0, 0, 1), (L-2, 2, 0, 0), (L-2, 1, 1, 0)$, $(L - 2, 1, 0, 1), (L - 2, 0, 2, 0), (L - 2, 0, 1, 1), (L - 2, 0, 0, 2)$, and so on. Thus, the process reduces the allocations sequentially through the nested loops by one-acre increments, adding that acre to the next crop in the ordering and re-evaluating the expected loss function at each trial allocation. Save the minimum expected loss found and the allocation associated with it; a simple if statement at the end of each loss function evaluation can be used to determine whether the new candidate allocation is an improvement on the current trial optimal solution. Note that even for a very small farm of 100 acres, such a grid search would consist of slightly more than one million trial allocations, each requiring the computation of B evaluations of the loss functions ($B = 25,000$ for Lence and Hayes). This would probably take three weeks on a reasonably fast (120MHz) PC according to a few trial programs used for rough approximation.

An alternative grid search routine would use a larger increment (or mesh) and then use a finer mesh in the area determined to be the neighborhood of the optimal allocation. For example, with 100 acres to allocate, one might start with increments of 10 acres. This would reduce the number of trial allocations to $11^3 = 1331$. A search of that size can be completed in approximately one hour. Then another search can be performed on the space of ± 5 acres on either side of the best allocation from the first grid search. This second search, even though the mesh is back down to one-acre increments has the same number of trial allocations, 1331, and can, therefore, be completed in an additional hour. Such an approach can be generalized to include as many repetitions with successively smaller grid increments as necessary to find an optimal solution with the desired level of numerical accuracy. This method will be preferred for applications with decision variables of large enough dimension (often 4 or greater) as to make a single grid search as described first too time-consuming.

The third possible approach is an active search algorithm that begins with an arbitrary trial allocation (decision) and then moves in successive steps toward new allocations that improve (decrease) the value of the expected loss (objective) function. This is completely analogous to iterative search algorithms used to solve nonlinear optimization problems (Gill, Murray, and Wright, 1985). The general approach is to begin with some initial trial decision, say an allocation of $(L/4, L/4, L/4, L/4)$, and then perturb it to see which reallocation most improves the decision. Thus, one would first evaluate the expected loss of the initial allocation. Then one could evaluate four potential new allocations of the form $(L/4 + 3, L/4 - 1, L/4 - 1, L/4 - 1)$ where each of the four crops is sequentially increased in acreage as the other three are minimally decreased. These four expected loss values would be examined and the one with the minimum expected loss would be chosen as the new trial allocation. Such a procedure would be repeated until such time as the optimal allocation is approached. Once close to the estimated optimal allocation, one might shift to a slightly more sophisticated approach in which 12 new allocations are considered, consisting of all combinations that add one acre to crop i and subtract one acre from crop j. Choosing the minimum expected loss decision from among these 12, with the caveat that it must have lower expected loss than the current trial allocation, one then proceeds cautiously toward the final, optimal allocation. When none of the 12 new allocations has lower expected loss than the current trial allocation, you have found the optimal acreage allocation given the numerical precision of your Monte Carlo sampling approximation and the search increments specified by your algorithm (here set at one acre).

Note that even an active search of 100 steps should be accomplished in an amount of time equivalent to the second type of grid search with changing mesh sizes. In fact, the active search method will generally be quicker in terms of computer time; however, it will take more time to write the computer program than for a grid search. Thus, one may choose the grid search except in cases where the dimension of the decision vector is very high, the search area is very wide, or when evaluation of the expected loss function is slow. In applications where finding the optimal decision may be difficult or extremely time-consuming, the researcher may want to do a small-scale time test of a few iterations of each method and/or evaluation of the expected loss function and then compute expected computing time for each of the three methods. In any iterative numerical algorithms on a computer it is always wise to time a small-scale example first and extrapolate the time to complete the task (even before performing a Monte Carlo sampling exercise, I always time 10 or 100 draws). This will help you decide the size of the empirical sample you can afford to draw (with the accompanying numerical precision) and ensure that you have the time resources to complete the project. Nothing is worse than killing a job that you are sure must be stuck in an infinite loop only to discover that it was working fine and almost done; you just misjudged how long the program would take to complete.

Lence and Hayes (1995) found the optimal Bayesian acreage allocations for two different crop allocation applications and compared the answers to those from three other methods: the plug-in approach which fails the certainty equivalence test because the loss function is not linear-quadratic; an unbiased, non-Bayesian analytical estimate of the optimal decision under estimation risk due to Chalfant, Collender, and Subramanian (1990); and an approximate Bayesian decision vector due to Brown (1979) that is exact for posterior distributions that follow the multivariate normal distribution and loss functions that are negative exponential. All three of these decision vectors can be solved for analytically. Lence and Hayes (1995) found that none of the three alternative methods produces good answers for either of two applications performed, even when the loss function is negative exponential and the posterior distribution is in the form of a multivariate Student-t distribution. This seems to be strong evidence that the full Bayesian decision theory approach is worthwhile and should be performed more often in empirical applications.

If one suspects that the optimal allocation of activities (crops) can include some percentage of doing nothing (land laying fallow, a factory operating below capacity), one can easily include such an activity in the decision vector. Create an additional activity (crop) that has a "random" profit with a degenerate posterior distribution that has mean profit zero and a zero variance (and covariance with any alternative activities). This generalization of the problem solved by Lence and Hayes (1995) is straightforward and worth considering in any situation with probabilities of negative profits from some activities and a risk-averse (convex) loss function.

Other Possible Decision Theory Applications

Almost any economic decision problem relies on a set of parameter values that are unknown in empirical applications. All such problems are candidates for the Bayesian approach. However, to get the most benefit out of accounting for the estimation risk incurred by working with estimated parameters, one should be careful that the decision problem is such that estimation risk is important.

First, the expected loss function must be nonquadratic or the posterior distribution of the parameters must be significantly skewed. In most cases, that occurs in economic problems through the introduction of risk aversion. If the decision maker is risk-neutral, many reasonable objective functions will be linear-quadratic in the decision variables and the unknown parameters, resulting in the certainty equivalence principle holding and the Bayesian approach yielding the same answer as the plug-in approach of using the posterior mean estimates of the parameters as if they were deterministic. Second, the estimation risk must be economically significant. If the posterior distribution is highly concentrated around the mean, so that the ratio of posterior mean to posterior standard deviation is large (in absolute value), the proper treatment of estimation risk may not change the decision by an economically significant amount. For example, if the ratio of the absolute value of the posterior mean to posterior standard deviation is near or above 10 for all unknown parameters in a problem, taking the Bayesian decision theory approach will generally have virtually no effect on the estimated optimal decision.

Having satisfied yourself that the problem at hand is worth tackling through the Bayesian decision theory approach, a huge set of problems can be tackled through the general method outlined earlier for the case of acreage allocation. Some particularly well-suited situations that come to mind are listed soon. All that is needed to solve such problems are the insertion of a suitable loss function in place of the one used earlier (which is, in fact, often suitable) and the substitution of the correct posterior distribution of the unknown parameters. In many applications where informative priors may be appropriate, the posterior distribution will not have the Student-t form, even after marginalizing the variance components; therefore, many applications will have to rely on importance sampling instead of Monte Carlo sampling. One simply evaluates the expected loss function using the formula for importance sample (i.e., weighted average) instead of the simple average used with Monte Carlo sampling and proceeds identically to the preceding outline otherwise.

One potential type of application is determining the optimal storage quantity for a commodity that is either produced infrequently (many agricultural products) or with great temporal variation (again, mostly agricultural examples). Here, unknown parameters can be found in the supply and demand curves that control the decision and in the new supply to be forthcoming in future periods. This last parameter often has a sizeable estimation risk, which one would not want to ignore. Risk-averse marketing agents who hold the supplies comprise the only assumption left to make this problem fit the Bayesian decision theory paradigm.

Hedging decisions are another natural application of these techniques. Such an application was performed by Lence and Hayes (1994). A farmer or consumer of a commodity that has an associated actively traded future contract can diminish the profit risk (on either the revenue or cost side, respectively) inherent in their businesses by hedging in the futures markets. Here the unknown parameters are associated with demand parameters, production quantity (yield risk), and the difference between spot and futures prices (basis risk).

Life insurance underwriting would provide an interesting application. The underwriting company is trying to construct a portfolio of insurance (who it accepts or rejects) that will make it a profit in the long-run but must rely on extrapolations of life expectancies from tables based on previously observed data (because those buying the insurance come from a population whose true life expectancy cannot possibly be known at the time of purchase). If a suitable nonquadratic loss function could be constructed, this problem would be very revealing when compared to actuarial insurance rates and the underwriters observed behavior.

Many investment problems are well-suited to this sort of analysis, as long as the firm being studied can be postulated to have a nonquadratic objective function (be risk-averse or risk-taking). Unknown parameters can relate to the final product's demand, the production of the product, and the potential entry of new competitors. Given the difficulty in the investment literature of estimating dynamic models with adjustment costs with much empirical precision, the size of the estimation risk present makes the application of the Bayesian decision theory approach worthwhile.

Optimal Decisions for Dynamic Problems

Finally, in the area of resource economics, the optimal regulation of fisheries seems tailor-made for this approach, but a slight adaptation must be made to the solution technique. The loss function is definitely nonquadratic as the objective is generally to maximize the present value of the flow of profits from the fishery with some adjustment to ensure that the species is not fished to extinction (even if that strategy has a higher discounted profit flow). Estimation risk certainly exists because the fish population itself is an unknown random variable and the population growth process must also be estimated and entered into the decision problem as a set of constraints.

This problem falls into the realm of stochastic optimal control but can be attacked by the Bayesian decision science approach for discrete time problems that can be solved analytically conditional on the parameters (this is a fairly large class of problems that extends well beyond the standard linear-quadratic control problem). When such problems are solved by the classical calculus of variations approach (or Bellman's dynamic programming approach), the parameters are all taken as deterministic and a solution is found for the time path of the decision variables (often just a single decision variable). One cannot find the Bayesian solution by generating a large number of draws from the posterior distribution of the unknown parameters, solving for the optimal decision path through time conditional on each draw, and then taking an average of the decision paths. That is not the minimum expected loss decision path.

Instead, one must evaluate a set of candidate decision paths at each of the generated values from the parameters' posterior distribution, find the expected loss for each of the candidate decision paths, and choose the minimum expected loss decision path. Thus, instead of having to perform sequential analytical solutions conditional on some large number of possible parameter values (which would have made the problem impractable), one must simply find the expression for the value of the loss function (objective function) given a particular decision path. In general, such an expression can be found for most dynamic optimization problems with a form that is analytical and can be placed into a computer program.

Treating this expression as the loss function $F()$ described in the acreage allocation example, the only detail left is whether to grid search over decision paths or to perform an active search. In many cases, even though the problem must be converted to discrete time periods, the number of periods and number of possible decision variable values at each point in time will be such that a grid search will be infeasible. For a 20-period problem, even two possible decisions at each point in time would lead to a problem that might take more than a week to run on a pretty fast computer. Because most interesting dynamic problems must be run for at least 10 or 20 periods, an active search algorithm appears to be the best approach. Because many of the decision variables considered in such problems are not linked by adding up constraints as the acreage allocations were, some modification of the search approach is in order.

A simple approach would be to first solve the problem by a standard method using the posterior means of the unknown parameters as if they were deterministic. Use this decision path as an initial candidate optimal path. Assume the decision path consists of values for m control variables. Then evaluate the expected loss for this initial path along with m perturbed paths, each consisting of the initial candidate path with a single decision variable perturbed upward by a predetermined increment. These increments will be important as they must be small enough not to jump over optimal solutions but not so small as to cause huge numbers of steps to be necessary to reach the final optimum. Find the perturbation that causes the largest change in expected loss. If the change lowers the expected loss, increase the value of that decision variable by the increment and repeat the process; if the change increases expected loss, decrease the value of the decision variable by the increment and repeat the process. At all steps, check to make sure that the new candidate decision path has a lower expected loss than the previous one. If not, discard that candidate, and then use smaller increments, perturb a different decision variable, or check to see whether the optimal decision path has been reached. Gill, Murray, and Wright (1985) can be consulted for good suggestions on both search algorithms and determining when the optimal decision has been found.

When the search is completed, you have an optimal decision path that accounts for the parameter uncertainty. No derivatives need be taken within the computer program (some may have to be taken by the researcher to derive either the loss function or the initial candidate decision path). Further, even for fairly complex loss functions, on the order of 20 decision variables, and a search algorithm that takes several thousand steps before finding the minimum expected loss decision, such a program should be capable of completion on an average PC overnight (i.e., in less than 15 hours). Thus, the problem is not overly time-consuming in terms of computing.

The payoff to such an approach to dynamic optimization problems could be quite large. First, policy makers could evaluate the benefits to conducting research that shrinks the parameter uncertainty in their decision problems. Without taking the Bayesian approach, parameter uncertainty is ignored and the benefits of improving our knowledge of the underlying parameters cannot be calculated. Second, one can calculate important policy-relevant figures such as the probability of extinction under the optimal decision path and other alternative paths proposed by policy makers or other agents involved in the industry (fishermen, environmental groups, etc.). Finally, one could place a term in the loss function to severely penalize any decision path that had a probability of extinction that exceeded a preselected limit. If the penalty is made large enough, no path that exceeds the limit for extinction probability will be selected, effectively imposing a chance constraint on the decision problem. Such chance constraints have many applications in other types of dynamic and static optimization problems and are much easier to implement in the Bayesian decision theory approach than under classical solution techniques.

References

Berger, J. O. (1985). *Statistical Decision Theory and Bayesian Analysis*. New York: Springer-Verlag.

Brown, S. J. (1979). "Optimal portfolio choice under uncertainty: A Bayesian approach," *Estimation Risk and Optimal Portfolio Choice*, eds. V. Bawa, S. Brown, and R. Klein. Amsterdam: North-Holland.

Chalfant, J. A., R. N. Collender, and S. Subramanian (1990). "The mean and variance of the mean-variance decision rule." *American Journal of Agricultural Economics* 72, 966-974.

Gill, P. E., W. Murray, and M. H. Wright (1985). *Practical Optimization*. New York: Academic Press.

Lence, S. H., and D. J. Hayes (1994). "Parameter-based decision making under estimation risk: An application to futures trading." *Journal of Finance* 49, 345-357.

Lence, S. H., and D. J. Hayes (1995). "Land allocation in the presence of estimation risk." *Journal of Agricultural and Resource Economics* 20, 49-63.

Bibliography

Key Theory Details

Bawa, V., S. Brown, and R. Klein, Eds. (1979). *Estimation Risk and Optimal Portfolio Choice*. Amsterdam: North-Holland.

Berger, J. O. (1985). *Statistical Decision Theory and Bayesian Analysis*. New York: Springer-Verlag.

Box, G. E. P., and G. C. Tiao (1973). *Bayesian Inference in Statistical Inference*. Reading, Mass.: Addison-Wesley. Now also New York: Wiley (1992).

Casella, G., and E. I. George (1992). "Explaining the Gibbs sampler." *The American Statistician* 46, 167–174.

Chan, K. S. (1993). "Asymptotic behavior of the Gibbs sampler." *Journal of the American Statistical Association* 88, 320–326.

Chib, S. (1995). "Marginal likelihood from the Gibbs output." *Journal of the American Statistical Association* 90, 1313–1321.

Chib, S., and E. Greenberg (1995). "Hierarchical analysis of SUR models with extensions to correlated serial errors and time-varying parameter models." *Journal of Econometrics* 68, 339–360.

Drèze, J., and J.-F. Richard (1983). "Bayesian analysis of simultaneous equation systems." *Handbook of Econometrics, vol. 1*, eds. Z. Griliches and M. D. Intriligator. New York: North Holland.

Geman, S., and D. Geman (1984). Stochastic relaxation, Gibbs distributions, and Bayesian restoration of images. *IEEE Trans. Pattern Anal. and Machine Intelligence* 6, 721–741.

Geweke, J. (1986). "Exact inference in the inequality constrained normal linear regression model." *Journal of Applied Econometrics* 1, 127–141.

Geweke, J. (1988). "Antithetic acceleration of Monte Carlo integration in Bayesian inference." *Journal of Econometrics* 38, 73–89.

98   Bibliography

Geweke, J. (1989). "Bayesian inference in econometric models using Monte Carlo integration." *Econometrica* 57, 1317–1339.

Geweke, J. (1991). Generic, algorithmic approaches to Monte Carlo integration in Bayesian inference. *Contemporary Mathematics vol.115*, eds. N. Flournoy and R. K. Tsutakawa. Providence: American Mathematical Society.

Geweke, J. (1995). "Monte Carlo simulation and numerical integration." Federal Reserve Bank of Minneapolis Research Dept. Staff Report 192.

Geyer, C. (1992). "Practical Markov chain Monte Carlo." *Statistical Science* 7, 473–482.

Good, I. J. (1965). *The Estimation of Probabilities: An Essay on Modern Bayesian Methods*. Cambridge, Mass.: MIT Press.

Hammersley, J. M., and D. C. Handscomb (1964). *Monte Carlo Methods*. London: Chapman and Hall.

Judge, G. G., R. C. Hill, and M. E. Bock. "An adaptive empirical Bayes estimator of the multivariate normal mean under quadratic loss." *Journal of Econometrics* 44, 189–213.

Kadane, J. B. (1984). *Robustness of Bayesian Analysis*. Amsterdam: North-Holland.

Kloek, T., and H. K. van Dijk (1978). "Bayesian estimates of simultaneous equation system parameters: An application of integration by Monte Carlo." *Econometrica* 46, 1–19.

Leamer, E. E. (1978). *Specification Searches*. New York: Wiley.

McCulloch, R., and P. E. Rossi (1994). "An exact likelihood analysis of the multinomial probit model." *Journal of Econometrics* 64, 207–240.

Poirier, D. J. (1988). "Frequentist and subjectivist perspectives on the problems of model building in economics." *Journal of Economic Perspectives* 2, 121–144.

Poirier, D. J. (1994). "Jeffreys' prior for logit models." *Journal of Econometrics* 63, 327–339.

Pratt, J. W., H. Raiffa, and R. Schlaifer (1965). *Introduction to Statistical Decision Theory*, New York: McGraw–Hill Book Company.

Pratt, J. W., and R. Schlaifer (1988). "On the interpretation and observation of laws." *Journal of Econometrics* 39, 23–52.

Richard, J.-F. (1973). *Posterior and Predictive Densities for Simultaneous Equation Models*. Berlin: Springer-Verlag.

Schervish, M. J. (1995). *Theory of Statistics*. New York: Springer-Verlag.

Schervish, M. J., and B. P. Carlin (1992). "On the convergence of successive substitution sampling." *Journal of Computational and Graphical Statistics* 1, 111–127.

Smith, A. F. M., and A. E. Gelfand (1992). "Bayesian statistics without tears: A sampling–resampling perspective." *The American Statistician* 46, 84–88.

Tanner, M. A. (1996). *Tools for Statistical Inference: Methods for the Exploration of Posterior Distributions and Likelihood Functions, third edition.* New York: Springer–Verlag.

Tierney, L. (1994). "Markov chains for exploring posterior distribution (with discussion)." *Annals of Statistics* 22, 1701–1762.

Uhlig, H. (1994). "What macroeconomists should know about unit roots: A Bayesian perspective." *Econometric Theory* 10, 645–671.

West, M., and J. Harrison (1989). *Bayesian Forecasting and Dynamic Models.* New York: Springer-Verlag.

Winkler, R. L. (1980). "Prior information, predictive distribution, and Bayesian model building." *Bayesian Analysis in Econometrics and Statistics*, ed. A. Zellner. Amsterdam: North–Holland.

Zellner, A. (1971). *An Introduction to Bayesian Inference in Econometrics.* New York: John Wiley & Sons.

Zellner, A. (1988). "Bayesian analysis in econometrics." *Journal of Econometrics* 37, 27–50.

Zellner, A., and S. B. Park (1979). "Minimum expected loss (MELO) estimators for functions of parameters and structural coefficients of econometric models." *Journal of the American Statistical Association*, 74, 185–193.

Applications: Econometrics

Albert, J. H., and S. Chib (1993). "Bayes inference via Gibbs sampling of autoregressive time series subject to Markov mean and variance shifts." *Journal of Business & Economic Statistics* 11, 1–15.

Andrew, R. W., J. O. Berger, and M. H. Smith (1993). "Bayesian estimation of manufacturing effects in a fuel economy model." *Journal of Applied Econometrics* 8, S5-S18.

Atkinson, S. E., and T. D. Crocker (1987). "A Bayesian approach to assessing the robustness of hedonic property." *Journal of Applied Econometrics* 2, 27–45.

Barnett, G., R. Kohn, and S. Sheather (1996). "Bayesian estimation of an autoregressive model using Markov chain Monte Carlo." *Journal of Econometrics* 74, 237–254.

Bauwens, L., D. G. Fieberg, and M. F. J. Steel (1994). "Estimating end-use demand: A Bayesian approach." *Journal of Business & Economic Statistics* 12, 221–231.

Bauwens, L., W. Polasek, and H. K. van Dijk, Eds. (1996). "Bayes, Bernoullis and Basel." *Journal of Econometrics: Annals of Econometrics* 75, 1–238.

Berger, J., and T. Sellke (1987). "Testing of a point null hypothesis: The irreconcilability of significance levels and evidence." *Journal of the American Statistical Association* 82, 112–139.

Broeck, J. van den, G. Koop, J. Osiewalski, and M. F. J. Steel (1994). "Stochastic frontier models: A Bayesian perspective." *Journal of Econometrics* 61, 273–303.

Canova, F. (1992). "An alternative approach to modeling and forecasting seasonal time series." *Journal of Business & Economic Statistics* 10, 97–108.

Canova, F. (1993). "Forecasting time series with common seasonal patterns." *Journal of Econometrics* 55, 173–200.

Chalfant, J. A. (1993). "Estimation of demand systems using informative priors." *American Journal Agricultural Economic* 75, 1200–1205.

Chalfant, J. A., R. S. Gray, and K. J. White (1991). "Evaluating prior beliefs in a demand system: The case of meats demand in Canada." *American Journal Agricultural Economic* 73, 476–490.

Chib, S. (1992). "Bayes inference in the Tobit censored regression model." *Journal of Econometrics* 51, 79–99.

Chib, S. (1993). "Bayes regression with autoregressive errors: A Gibbs sampling approach." *Journal of Econometrics* 58, 275–294.

Chib, S. (1996). "Calculating posterior distributions and modal estimates in Markov mixture models." *Journal of Econometrics* 75, 79–97.

Chib, S., and E. Greenberg (1994). "Bayes inference in regression models with ARMA(p, q) errors." *Journal of Econometrics* 64, 183–206.

de Alba, E. (1988). "Disaggregation and forecasting." *Journal of Business & Economic Statistics* 6, 197–206.

DeJong, D. N. (1992). "Co-integration and trend-stationarity in macroeconomic time series: Evidence for the likelihood function." *Journal of Econometrics* 52, 347–370.

DeJong, D. N. (1993). "Bayesian inference in limited dependent variable models: An application to measuring strike duration." *Journal of Applied Econometrics* 8, 115–128.

DeJong, D. N., and B. F. Ingram, and C. H. Whiteman (1996). "A Bayesian approach to calibration." *Journal of Business & Economic Statistics* 14, 1–9.

DeJong, D. N., and C. H. Whiteman (1991). "Reconsidering 'trends and random walks in macroeconomic time series'." *Journal of Monetary Economics* 28, 221–254.

DeJong, D. N., and C. H. Whiteman (1991). "The case for trend-stationarity is stronger than we thought." *Journal of Applied Econometrics* 4, 413–421.

DeJong, D. N., and C. H. Whiteman (1991). "The temporal stability of dividends and stock prices: Evidence for the likelihood function." *American Economic Review* 81, 600–617.

DeJong, D. N., and C. H. Whiteman (1993). "Estimating moving average parameters: Classical pileups and Bayesian posteriors." *Journal of Business & Economic Statistics* 11, 311–317.

Doan, T., R. Litterman, and C. A. Sims (1984). "Forecasting and conditional projection using realistic prior distributions." *Econometric Reviews* 3, 1–100.

Dorfman, J. H. (1993). "Bayesian efficiency tests for commodity futures markets." *American Journal of Agricultural Economics* 75, 1206–1210.

Dorfman, J. H. (1995). "A numerical Bayesian test for cointegration of AR processes." *Journal of Econometrics* 66, 289–324.

Dorfman, J. H. (1996). "Modeling multiple adoption decisions in a joint framework." *American Journal of Agricultural Economics* 78, 547–557.

Dorfman J. H., and A. M. Havenner (1992). "A Bayesian approach to state space multivariate time series modeling." *Journal of Econometrics*, 52, 315–346.

Dorfman, J. H., and W. D. Lastrapes (1996). "The dynamic responses of crop and livestock prices to money-supply shocks: A Bayesian analysis using long-run identifying restrictions." *American Journal of Agricultural Economics* 78, 530–541.

Drèze, J. (1977). "Bayesian regression analysis using poly-t densities." *Journal of Econometrics* 6, 329–254.

Geweke, J. (1988). "The secular and cyclical behavior of real GDP in 19 OECD countries, 1957–1983." *Journal of Business & Economics Statistics* 6, 479–486.

Geweke, J. (1989). "Exact predictive densities for linear models with ARCH disturbances." *Journal of Econometrics* 40, 63–86.

Geweke, J. (1993). "Bayesian treatment of the independent Student-t linear model." *Journal of Applied Econometrics* 8, S19-S40.

Geweke, J. (1996). "Bayesian reduced rank regression in econometrics." *Journal of Econometrics* 75, 121–146.

Geweke, J., M. Keane, and D. Runkle (1994). "Alternative computational approaches to statistical inference in the multinomial probit model." *Review of Economics and Statistics* 76, 609–632.

Gilley, O. W., and R. K. Pace (1995). "Improving hedonic estimation with an inequality restricted estimator." *Review of Economics and Statistics* 77, 609–621.

Griffiths, W., and G. Judge (1992). "Testing and estimating location vectors when the error covariance matrix is unknown." *Journal of Econometrics* 54, 121–138.

Harrison, P. J., and C. F. Stevens (1976). "Bayesian forecasting." *Journal of the Royal Statistical Society B* 38, 205–247.

Hayes, D. J., T. I. Wahl, and G. W. Williams (1990). "Testing restrictions on a model of Japanese meat demand." *American Journal Agricultural Economic* 72, 556–566.

Hirschberg, J. G., and D. J. Slottje (1994). "An empirical Bayes approach to analyzing earnings functions for various occupations and industries." *Journal of Econometrics* 61, 65–79.

Hoek, H., A. Lucas, and H. K. van Dijk (1995). "Classical and Bayesian aspects of robust unit root inference." *Journal of Econometrics* 69, 27–59.

Hsiao, C., D. C. Mountain, and K. H. Illman (1995). "A Bayesian integration of end-use metering and conditional-demand analysis." *Journal of Business & Economic Statistics* 13, 315–326.

Inclan, C. "Detection of multiple changes of variance using posterior odds." *Journal of Business & Economic Statistics* 11, 289–300.

Jacquier, E., N. G. Polson, and P. E. Rossi (1994). "Bayesian analysis of stochastic volatility models," with discussion. *Journal of Business & Economic Statistics* 12, 371–389.

Kadane, J. B., N. H. Chan, and L. J. Wolfson (1996). "Priors for unit root models." *Journal of Econometrics* 75, 99–111.

Kato, H., S. Naniwa, M. Ishiguro (1996). "A Bayesian multivariate nonstationary time series model for estimating mutual relationships among variables." *Journal of Econometrics* 75, 147–161.

Kennedy, P., and D. Simons (1991). "Fighting the teflon factor: Comparing classical and Bayesian estimators for autocorrelated errors." *Journal of Econometrics* 48, 15–27.

Kleibergen, F., and H. K. van Dijk (1993). "Nonstationarity in GARCH models: A Bayesian analysis." *Journal of Applied Econometrics* 8, S41-S61.

Kooiman, P., H. K. van Dijk, and A. R. Thurik (1985). "Likelihood diagnostics and Bayesian analysis of a microeconomic disequilibrium model for retail services." *Journal of Econometrics* 29, 121–148.

Koop, G. (1991). "Cointegration tests in present value relationships: A Bayesian look at the bivariate properties of stock prices and dividends." *Journal of Econometrics* 49, 105–139.

Koop, G. (1991). "Intertemporal properties of real output: A Bayesian analysis." *Journal of Business & Economic Statistics* 9, 253–265.

Koop, G. (1992). "Aggregate shocks and macroeconomic fluctuations: A Bayesian approach." *Journal of Applied Econometrics* 7, 395–411.

Koop, G. (1992). "'Objective' Bayesian unit root tests." *Journal of Applied Econometrics* 7, 65–82.

Koop, G. (1994). "Bayesian semi-nonparametric ARCH models." *Review of Economics and Statistics* 76, 176–181.

Koop, G. (1996). "Parameter uncertainty and impulse response analysis" *Journal of Econometrics* 72, 135–149.

Koop, G., E. Ley, J. Osiewalski, and M. F. J. Steel (1997). "Bayesian analysis of long memory and persistence using ARFIMA models." *Journal of Econometrics* 76, 149–169.

Koop, G., J. Osiewalski, and M. F. J. Steel (1994). "Bayesian efficiency analysis with a flexible form: The AIM cost function." *Journal of Business & Economic Statistics* 12, 339–346.

Koop, G., J. Osiewalski, and M. F. J. Steel (1994). "Posterior properties of long-run impulse responses." *Journal of Business & Economic Statistics* 12, 489–492.

Koop, G., J. Osiewalski, and M. F. J. Steel (1995). "Bayesian long-run prediction in time series models." *Journal of Econometrics* 69, 61–80.

Koop, G., J. Osiewalski, and M. F. J. Steel (1997). "Bayesian efficiency analysis through individual effects: Hospital cost frontiers." *Journal of Econometrics* 76, 77–105.

Koop, G., and D. J. Poirier (1993). "Bayesian analysis of logit models using natural conjugate priors." *Journal of Econometrics* 56, 323–340.

Koop, G., and D. J. Poirier (1994). "Rank-ordered logit models: An empirical analysis of Ontario voter preferences." *Journal of Applied Econometrics* 9, 369–388.

Koop, G., and M. F. J. Steel (1994). "A decision theoretic analysis of the unit root hypothesis using mixtures of general elliptical models." *Journal of Business & Economic Statistics* 12, 93–106.

Leamer, E. E. (1973). "Multicollinearity: A Bayesian interpretation." *Review of Economics and Statistics* 55, 371–380.

Lee, J. C., and D. J. Sabavala (1987). "Bayesian estimation and prediction for the beta-binomial model." *Journal of Business & Economic Statistics* 5, 357–367.

LeSage, J. P. (1990). Forecasting turning points in metropolitan employment growth rates using Bayesian techniques. *Journal of Regional Sciences* 30, 533–538.

LeSage, J. P. (1991). "Analysis and development of leading indicators using a Bayesian turning–points approach." *Journal of Business & Economic Statistics* 9, 305–316.

LeSage, J. P. (1992). "Scoring the composite leading indicators: A Bayesian turning points approach." *Journal of Forecasting* 11, 35–46.

LeSage, J. P. (1993). "Spatial modeling of agricultural markets." *American Journal of Agricultural Economics*. 75, 1211–1216.

LeSage, J. P., and M. Magura (1990). "Using Bayesian techniques for data pooling in regional payroll forecasting." *Journal of Business & Economic Statistics* 8, 127–136.

LeSage, J. P. and M. Magura (1992). "A mixture-model approach to combining forecasts." *Journal of Business & Economic Statistics* 10, 445–452.

Li, D. T., and J. H. Dorfman (1995). "A robust approach to predicting fluctuations in state-level employment growth." *Journal of Regional Sciences* 35, 471–484.

Li, D. T., and J. H. Dorfman (1996). "Predicting turning points through the integration of multiple models." *Journal of Business & Economic Statistics* 14, 421–428.

Litterman, R. (1986). "Forecasting with Bayesian vector autoregressions-five years of experience." *Journal of Business & Economic Statistics* 4, 25–38.

Lubrano, M. (1995). "Testing for unit roots in a Bayesian framework." *Journal of Econometrics* 69, 81–109.

McCulloch, R., and P. E. Rossi (1991). "A Bayesian approach to testing the arbitrage pricing theory." *Journal of Econometrics* 49, 141–168.

McCulloch, R., and P. E. Rossi (1994). "An exact likelihood analysis of the multinomial probit model." *Journal of Econometrics* 64, 207–240.

Min, C. K., and A. Zellner (1993). "Bayesian and non–Bayesian methods for combining models and forecasts with applications to forecasting international growth rates." *Journal of Econometrics* 56, 89–118.

Monahan, J. F. (1983). "Fully Bayesian analysis of ARMA time series models." *Journal of Econometrics* 21, 307–331.

Moulton, B. R. (1991). "A Bayesian approach to regression selection and estimation, with application to a price index for radio services." *Journal of Econometrics* 49, 169–193.

Park, S. R., and J. K. Kwon (1995). "Rapid economic growth with increasing returns to scale and little or no productivity growth." *Review of Economics and Statistics* 77, 332–351.

Petit, L. I. (1992). "Bayes factors for outlier models using the device of imaginary observations." *Journal of the American Statistical Association* 87, 541–545.

Phillips, P. C. B. (1991). "To criticize the critics: An objective Bayesian analysis of stochastic trends." *Journal of Applied Econometrics* 6, 333–364.

Phillips, P. C. B. (1995). "Bayesian model selection and prediction with empirical applications." *Journal of Econometrics* 69, 289–331.

Poirier, D. J. (1991). "A Bayesian view of nominal money and real output through a new classical macroeconomic window." *Journal of Business & Economic Statistics* 9, 125–148.

Poirier, D. J. (1996). "A Bayesian analysis of nested logit models." *Journal of Econometrics* 75, 163–181.

Pole, A., M. West, and J. Harrison (1994). *Applied Bayesian Forecasting and Time Series Analysis*. New York: Chapman-Hall.

Raynauld, J., and J.-G. Simonato (1993). "Seasonal BVAR models: A search along some time domain priors." *Journal of Econometrics* 55, 203–229.

Richard, J.-F., and M. F. J. Steel (1988). "Bayesian analysis of systems of seemingly unrelated regression equations under a recursive extended natural conjugate prior density." *Journal of Econometrics* 38, 7–37.

Schotman, P. (1996). "A Bayesian approach to the empirical valuation of bond options." *Journal of Econometrics* 75, 183–215.

Schotman, P., and H. K. van Dijk (1991). "A Bayesian analysis of the unit root in real exchange rates." *Journal of Econometrics* 49, 195–238.

Schotman, P., and H. K. van Dijk (1991). "On Bayesian routes to unit roots." *Journal of Applied Econometrics* 6, 387–401.

Shively, T. S., and R. Kohn (1997) "A Bayesian approach to model selection in stochastic coefficient regression models and structural time series." *Journal of Econometrics* 76, 39–52.

Sims, C. (1988). "Bayesian skepticism on unit root econometrics." *Journal of Economic Dynamics and Control* 12, 463–474.

Sims, C. A., and H. Uhlig (1992). "Understanding unit rooters: A helicopter tour." *Econometrica* 60, 121–124.

Smith, M., and R. Kohn (1996). "Nonparametric regression using Bayesian variable selection." *Journal of Econometrics* 75, 317–343.

Steel, M. F. J. (1991). "A Bayesian analysis of simultaneous equation models by combining recursive analytical numerical approaches." *Journal of Econometrics* 48, 83–117.

Steel, M. F. J., and J.-F. Richard (1991). "Bayesian multivariate exogeneity analysis: An application to a U.K. money demand equation." *Journal of Econometrics* 49, 239–274.

Thompson, P. A., and R. B. Miller (1986). "Sampling the future: A Bayesian approach to forecasting from univariate time series models." *Journal of Business & Economic Statistics* 4, 427–436.

Tsurumi, H., and H. Wago (1991). "Mean squared errors of forecast for selecting nonnested linear models and comparison with other criteria." *Journal of Econometrics* 48, 215–240.

Zellner, A. (1978). "Estimation of functions of populations means and regression coefficients: A minimum expected loss approach." *Journal of Econometrics* 8, 127–158.

Zellner, A., L. Bauwens, and H. K. van Dijk (1988). "Bayesian specification analysis and estimation of simultaneous equation models using Monte Carlo methods." *Journal of Econometrics* 38, 39–72.

Zellner, A., and R. Highfield (1988). "Calculation of maximum entropy distributions and approximation of marginal posterior distributions." *Journal of Econometrics* 37, 195–209.

Zellner, A., and C. Hong (1988). "Bayesian methods for forecasting turning points in economic time series: Sensitivity of forecasts to asymmetry of loss structures." *Leading Economic Indicators: New Approaches and Forecasting Records*, eds. K. Lahiri and G. Moore. Cambridge: Cambridge University Press.

Zellner, A., and C. Hong (1989). "Forecasting international growth rates using Bayesian shrinkage and other procedures." *Journal of Econometrics* 40, 183–202.

Zellner, A., C. Hong, and G. M. Gulati (1990). "Turning points in economic time series, loss structures and Bayesian forecasting." *Bayesian and Likelihood Methods in Statistics and Econometrics: Essays in Honor of George A. Barnard*, eds. S. Geisser, J. Hodges, S. J. Press, and A. Zellner, Amsterdam: North Holland, pp. 371–393.

Zellner, A., C. Hong, and C. Min (1991). "Forecasting turning points in international output growth rates using Bayesian exponentially weighted autoregression, time–varying parameter, and pooling techniques." *Journal of Econometrics* 49, 275–304.

Zellner, A., and P. E. Rossi (1984). "Bayesian analysis of dichotomous quantal response models." *Journal of Econometrics* 25, 365–393.

Zellner, A., and A. Siow (1980). "Posterior odds ratios for selected regression hypotheses." *Bayesian Statistics*, eds. J. M. Bernado, M. H. DeGroot, D. V. Lindley, and A. F. M. Smith. Valencia: University Press.

Applications: Economic Decision Theory

Berger, J. (1984). "Bayesian input in Stein estimation and a new minimax empirical Bayes estimator." *Journal of Econometrics* 25, 87–108.

Brown, S. J. (1979). "Optimal portfolio choice under uncertainty: a Bayesian approach." *Estimation Risk and Optimal Portfolio Choice*, eds. V. Bawa, S. Brown, and R. Klein. Amsterdam: North-Holland.

Ethridge, D. E., P. Zhang, B. E. Dahl, R. T. Ervin, and J. Rushemeza (1990). "Cattle ranching production and marketing strategies under combined price and weather risks." *Western Journal of Agricultural Economics* 15, 175–185.

Klein, R. W., L. C. Rafsky, D. S. Sibley, and R. D. Willig (1978). "Decisions with estimation uncertainty." *Econometrica* 46, 1363–1387.

Lence, S. H., and D. J. Hayes (1994). "Parameter-based decision making under estimation risk: An application to futures trading." *Journal of Finance* 49, 345–357.

Lence, S. H., and D. J. Hayes (1994). "The empirical minimum-variance hedge." *American Journal Agricultural Economics* 76, 94–104.

Lence, S. H., and D. J. Hayes (1995). "Land allocation in the presence of estimation risk." *Journal of Agricultural & Resource Economics* 20, 49–63.

Index